D0116757

World Beneath the Sea

Prepared by the Special Publications Division
National Geographic Society, Washington, D. C.

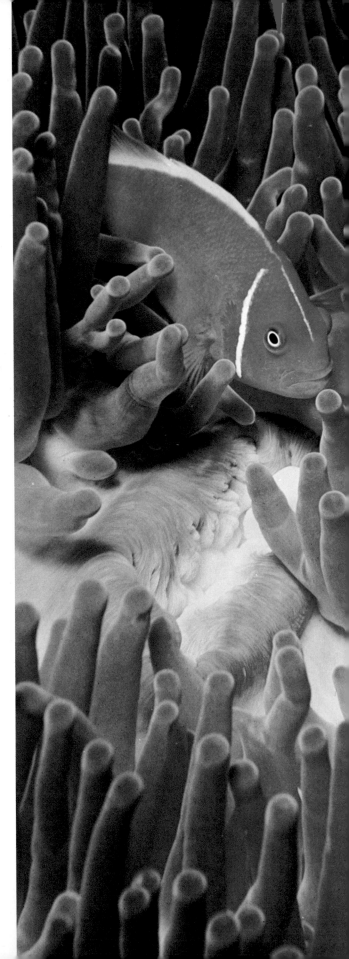

WORLD BENEATH THE SEA
Contributing Authors
BILL BARADA, GEORGE F. BASS, R. FRANK
   BUSBY, ROBERT C. COWEN, JAMES DUGAN,
   LUIS MARDEN, CYNTHIA RIGGS STOERTZ

Published by
THE NATIONAL GEOGRAPHIC SOCIETY
MELVIN M. PAYNE, *President*
MELVILLE BELL GROSVENOR, *Editor-in-Chief*
GILBERT M. GROSVENOR, *Editor*

Prepared by
THE SPECIAL PUBLICATIONS DIVISION
ROBERT L. BREEDEN, *Editor*
DONALD J. CRUMP, *Associate Editor*
PHILIP B. SILCOTT, *Senior Editor*
TADD FISHER, *Project Editor*
THEODORE S. AMUSSEN, DAVID R. BRIDGE,
   MARY ANN HARRELL, GERALD S. SNYDER,
   Assistant Project Editors
JOHANNA G. FARREN, *Research and Style*
LOUISE GRAVES, TEE LOFTIN SNELL, JANE
   STEIN, PEGGY WINSTON, *Research*
LINDA M. SEEMAN, *Illustrations Research*
WAYNE BARRETT, RONALD M. FISHER, LOUISE
   GRAVES, WILLIAM R. GRAY, JR., DAVID
   MALLORY, GERALD S. SNYDER, *Picture Legends*
LUBA BALKO, MARGARET S. DEAN, CHRISTINE
   J. SCHWARTZ, *Production-Editorial Assistants*

Illustrations and Design
DAVID R. BRIDGE, *Picture Editor*
JOSEPH A. TANEY, *Art Director*
JOSEPHINE B. BOLT, *Assistant Art Director*

Production and Printing
RONALD M. FISHER, *Production*
JAMES R. WHITNEY, *Engraving and Printing*
JOHN R. METCALFE, *Assistant,*
   Engraving and Printing

Revision Staff
MERRILL WINDSOR, *Managing Editor;* DR.
   ROBERT B. ABEL, *Consultant;* LEON M.
   LARSON, ARTHUR P. MILLER, CYNTHIA
   RIGGS STOERTZ, *Manuscript Editors;* RONALD
   M. FISHER, *Assistant Editor;* DAVID R.
   BRIDGE, *Picture Editor;* JILL R. DURRANCE,
   MARION K. INGERSOLL, JANE R. McCAULEY,
   LINDA LU MOORE, ANN CROUCH RESH,
   TEE LOFTIN SNELL, CYNTHIA RIGGS
   STOERTZ, *Research;* WILLIAM R. GRAY,
   H. ROBERT MORRISON, *Picture Legends;*
   URSULA PERRIN, *Design;* ROBERT W.
   MESSER, *Production Manager;* GEORGE V.
   WHITE, *Assistant Production Manager;* MARY
   L. BERNARD, RAJA D. MURSHED, MARGARET
   MURIN SKEKEL, *Production Assistants;* JOHN
   R. METCALFE, *Engraving and Printing;*
   JOLENE McCOY, *Index*

Second Edition 1973

*Living among stinging tentacles of a sea anemone, a
three-inch skunk clownfish attracts larger prey to
share with its host. Overleaf: Sunlit air bubbles dance
past a branch of elkhorn coral as a diver off the Florida coast enters the captivating world beneath the sea.*

CHAMBERED NAUTILUS, APPROXIMATELY 1/4 LIFE-SIZE (PAGE ONE)
DOUGLAS FAULKNER, PAGE ONE AND RIGHT; JERRY GREENBERG, OVERLEAF

Foreword

AT LONG LAST, man has begun intensive exploration of his underwater world. The oceans, despite their great hazards, beguile him with their extravagant bounty. For centuries man probed the mysteries of the sea, venturing upon it in frail ships or exploring its shallows in crude diving machines. But only in the past four decades has he begun to study the depths with his own eyes. From coral reefs, and even down to the dark abysses, he is uncovering secrets nature has guarded for millions of years.

As one addicted to the sea, I can attest that a diver's attraction to the ocean floor is even more compelling than the sailor's love for open water. To this day, more than a decade later, I vividly recall diving into the Aegean Sea to pry loose some fragments of red pottery I had spotted from the surface. I retrieved no gold, no statuary, only barnacle-encrusted shards, but I carried away the most rewarding prize of all—the excitement of discovery.

You, as members of the Society, have shared this excitement of exploring the unknown seas through the pages of the NATIONAL GEOGRAPHIC. Since 1888, the Society has published numerous accounts of many adventurous men of this ocean era—men like Dr. William Beebe, first naturalist to enter the abyss; Jacques-Yves Cousteau, the intrepid French pioneer of the Aqua-Lung; Dr. Harold E. Edgerton, developer of deep-sea cameras and the electronic flash; Edwin A. Link, inventor and undersea explorer; George F. Bass and Peter Throckmorton, underwater archeologists. With the Society's support, they have won major victories in man's conquest of the mysterious sea.

When the Society decided to publish this book about the earth's last frontier, one author came immediately to mind: James Dugan, perhaps the world's foremost chronicler of the undersea world and its heroes. Tragically, before Jim could complete his manuscript for the first edition of this volume, he suffered a heart attack and died at sea on June 1, 1967. Although it cannot console those of us he left behind, Jim died doing what he cherished most: writing about the sea he loved.

For this revised edition of *World Beneath the Sea*, other writers have taken up the challenge defined by Jim for all men of the sea: "to expand human knowledge and wonder." It is a challenge worthy of our age, and luckily there still exist men and women—in the tradition of James Dugan—eager to confront it.

Gilbert M. Grosvenor

Contents

*In a coral glade off Isla de Cozumel, Mexico, an
amateur diver 70 feet down gathers swaying sea fans.*
BOB HOLLIS

1

Exploring the Ocean World

BY JAMES DUGAN

Pacific breaker explodes in a geyser of water against the coast of Hawaii. Earth's restless oceans—actually one vast interconnected sea—flood seven-tenths of the globe. For centuries man knew little more than the surface; now he has begun to explore the innermost depths.

INSIDE THE MINIATURE SUBMARINE, we lay on our stomachs peering through portholes into a slowly darkening world. As the Mediterranean Sea closed over us, the light dimmed from golden on the sunlit surface to aquamarine, and then to smoky green.

"Diving Saucer Mission number 402, October 4, 1965," my companion reported to the tape recorder logging our voyage. "Pilot, Albert Falco, *Observateur,* Jimmy Dugan. *Destination, Précontinent Trois.*"

Falco steered a roundabout course to a submerged outpost—Jacques-Yves Cousteau's Continental Shelf Station (Conshelf) Three—328 feet down, off Cap Ferrat on the coast of southern France. In that sea-floor house six of my friends had lived and worked for 12 days without surfacing.

"Oceanauts," Cousteau dubbed this new breed of undersea technicians, members of a small international vanguard pioneering the occupation of the continental shelves. Their historic efforts—requiring submission to unremitting cold and dependence upon exotic breathing-gas mixtures to keep alive in an environment inhospitable to earthmen—held great significance for the future.

The gently sloping continental shelves are the pedestals of the land masses, almost equal to Africa in total area and incredibly rich in foods, fuels, and minerals. Most important, on an average their outer edges lie only a little more than 600 feet beneath the surface, a depth well within reach of underwater explorers. The oceanauts were helping to perfect the kind of sea-floor residence that may soon shelter marine miners and farmers. I couldn't help comparing my Conshelf Three visit to dropping in on a group of hopeful, isolated California prospectors in 1850.

As Falco and I descended, I had no sense of movement and became engrossed in checking the depth gauge...50 feet... 100...150. It grew darker outside. At first glimpse the nether slope of Cap Ferrat seemed an illusion. Against its vague shape, I saw hundreds of white specks, tiny organisms, drifting upward outside the ports.

Locked into a small, snug planet, we

9

JAMES M. ROBINSON (ABOVE) AND MARC JASINSKI

Drifting freely and weightlessly, a diver explores sheer walls of a submarine canyon off the coast of Tunisia. Beyond him stretches earth's last real frontier—a dark three-dimensional wilderness of canyon deeps, abyssal plains, and enormous mountain ranges virtually unknown to man. At the threshold of the deep frontier an underwater adventurer (left) roams with camera, plastic sketch pad, and spear. Delicately hued reefs like this one in Hawaiian waters lure growing numbers of amateur divers into fantastic coral realms that begin only a few feet beneath the surface.

coasted through a marine Milky Way. As I lolled beside the imperturbable master of the three-dimensional voyage, I realized a wonderfully happy moment in my life. Only the whisper of the air-circulating system broke the silence. The extra pinch of oxygen doled to us increased my euphoria.

The sea had long been my companion, but never so intimately as now. I had spent much of my life sailing on it, writing about

THE AUTHOR: *"The sea imposes a modesty on those who try her,"* wrote James Dugan in 1956, and when he died on June 1, 1967, while working on this book, the sea's mysteries still captivated him. In 1944 he met Capt. Jacques-Yves Cousteau and began a collaboration that produced The Living Sea. *He also wrote the narration for Cousteau's two Oscar-winning films,* The Silent World *and* World Without Sun. *His other books:* The Great Iron Ship, Man Under the Sea, American Viking, *and* The Great Mutiny.

it, peering into it, prowling its beaches, and once, protestingly, I probed its restless shallows when Falco coaxed me to put on an Aqua-Lung. Now I had become part of this liquid biospace that bulks three hundred times larger than our accustomed terrestrial environment.

Of all the planets in our solar system, earth alone is known to contain oceans—and earth's oceans are actually one vast sea. This water nourishes life; without it, we would quickly perish. Fossils more than two and a half billion years old indicate that from the womb of the salty sea emerged the beginnings of all life on earth.

This world ocean, this vast culture broth and spring of life, blankets seven-tenths of the globe and shrouds awesomely spectacular topography. Soundings and electronic magic, rather than personal inspection, have revealed most of the little we know about the grandeur of the depths.

I saw how these revelations came about while cruising with Cousteau on his oceanographic research ship *Calypso* during one of his Indian Ocean expeditions. I remember how the ship steamed along, singing the song of sonar—high-pitched pings strung in hypnotic series, each note projected toward the bottom of the sea and caught on the rebound. The time interval between ping and echo measured the distance the sound had passed through water. Fascinated, I watched the recording sonar as its stylus mysteriously sketched on graph paper the contour of the slope 2,700 feet down.

But these sketches form mere shadows

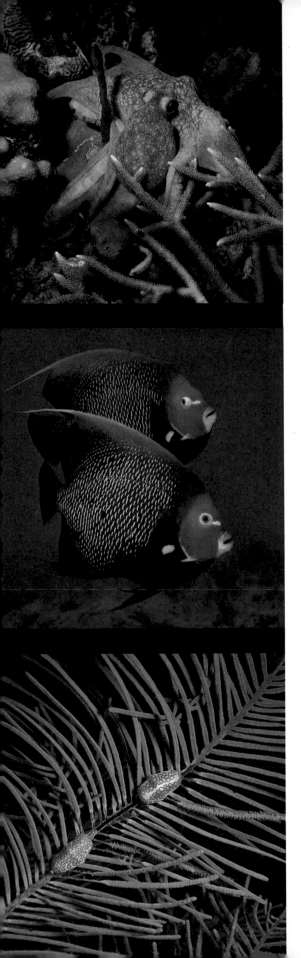

of reality. Most of the underwater canyons, trenches, valleys, mountains, and plains remain a territory not yet invaded by human explorers. And what a territory! Falco and I now lingered on the edge of this vast undersea world that covers more than 139 million square miles and contains superlative geography.

Earth's longest mountain range, the Mid-Oceanic Ridge, meanders between the continental land masses for 35,000 miles through all the oceans. Rising 6,000 to 12,000 feet above the bottom, the pinnacles of the ridge occasionally break through the surface to form islands, such as the Azores and Ascension. In expanse, the ridge almost matches the area of all the continents.

Although 29,028-foot Mount Everest is the roof of the land above the water, earth's greatest known height has its roots in the Pacific off Peru. There the Andes ascend 25,000 feet before surfacing, then climb another 23,000 feet—a total rise of more than nine miles.

The deep trenches claim my admiration most of all, for they have no rivals on dry land. In 1960, the bathyscaph *Trieste* carried two observers down nearly seven miles into the Mariana Trench in the Pacific— the farthest penetration made by man into the deepest known place in the ocean.

Unfortunately, bathyscaphs lack mobility; they serve best as marine elevators. And the more mobile conventional submarines are not designed for reaching great depths. Considering these facts, I realized I was reclining in a 21st-century sea craft. Already, descendants of the

Filming marvels of a coral reef off the Florida Keys, a photographer uses a ring-shaped flash reflector around the lens for close-ups of a convoluted brain coral. Tiny wrasses flit past unafraid. Its tentacles draped on staghorn coral, a briar octopus lurks to grab mollusks and crustaceans. French angelfish bear sharp cheek barbs that can stab an attacker. Spotted flamingo tongues graze on an undulating sea fan.

Lacy red sea fan (left), a colonial animal, sieves plank-ton through myriad tentacles. A vase sponge (above) draws food and oxygen through fine pores in its skin.

Swift, ferocious hunters, blackbarred jacks (above) school in warm seas. Mollusk without a shell, the Span-ish Dancer sea slug (below) glides above shelf coral.

highly maneuverable diving saucer were performing tasks far in excess of her depth capacity of 1,100 feet.

Shortly after Cousteau tested his first saucer in 1959, the overjoyed inventor wrote: "... already I can see that this odd jet-propelled vehicle will let us fulfill a dream: to descend deeper and stay longer than the free diver can, while still being able to move, look about, and even pick things up. This is revolutionary. The way opens for geological and biological research in a marine twilight zone no man could explore freely before."

Now Falco and I were roaming that zone, that mysterious realm "teeming, beckoning, unexplored," as Cousteau had put it. Our saucer could make little more than half a knot, but the clamlike craft glided about almost as easily as a fish. Two battery-powered hydrojet nozzles propelled us and controlled our direction. When Falco pointed the nozzles down, jets of water thrust the saucer up. When he pointed them up, we went down. To turn left, he used the power of the right nozzle only. To turn right, he used the left nozzle.

Actually, Falco had let the craft drift downward without power until we reached 270 feet. Then he turned on the jets and began putting the saucer through her acrobatic paces as he unerringly flitted about in the dark, prolonging our excursion.

Intruder in a strange animal kingdom crouches behind a jagged outcrop of living coral. Beside her sprout a vase sponge and a branching sea whip. Long mistaken for plants and stones, corals grow in all the seas, but reef-building varieties flourish only in tropic shallows. Cellulose-like stalked tunicates cluster in a colony (center). Anchoring themselves to twigs of coral or submerged rocks, these twin-mouthed creatures filter food from water drawn into one siphon and squirted out of the other. The sea anemone (bottom) often paralyzes fish with stinging tentacles, then draws its prey into its saclike body. With its false eye, the golden longnosed butterfly fish confuses attackers that find it hard to tell whether their quarry is coming or going.

He flipped a switch and the submarine's exterior lights blazed a trail through the water. The sea came alive. I saw a circus of color: red gorgonians, yellow sponges, and purple sea urchins on greenish-gray corals. Small fish pranced unafraid in the sudden brightness. An octopus sprawled on a sandy ridge. And all around us swarmed a host of tiny creatures—jellyfish, worms, crustaceans, and the young of crabs, lobsters, and fishes.

Falco gaily revolved the jets to swoop and climb. Then he reversed the flow to slow down and we hovered beside the reef. Cousteau had told me that despite the saucer's grunting and whining when under power, it did not frighten fish away, but attracted them instead. I wondered if they saw the submersible as an engaging clown of a clam with staring, curious eyes, a great yellow clam that refused to stay put on the bottom where it belonged.

We zoomed downward and began to glide over the bottom. I was completely disoriented. I guessed our direction as northwest, but the compass needle indicated east by southeast. Falco pressed on as confidently as a householder in the dusk of his own backyard.

"Regardez la grosse langouste," he said. I didn't see any big lobster. Seconds later it came into the arena of light, waving long antennae. My pilot had seen it beyond in the gloom. Turning to starboard, Falco

DAVID R. BRIDGE, NATIONAL GEOGRAPHIC STAFF

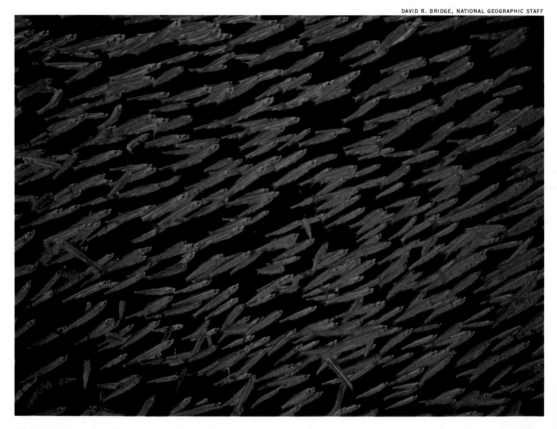

Shimmering cloud of reef silversides clusters above a wreck in Florida's John Pennekamp Coral Reef State Park. At right, the ocean beds drained of water emerge as a spectacular region unrivaled by anything on land. Towering peaks, plunging canyons, and plains flat as tabletops remain undisturbed by erosion of wind and rain.

Ships' echo sounders mapped this underwater world. The Mid-Oceanic Ridge, criss-crossed with fracture zones and grooved by a rift valley, winds for 35,000 miles through all of earth's oceans. On either side, broad basins and plains stretch to the shallow continental shelves.

PAINTING BY HEINRICH C. BERANN BASED ON BATHYMETRIC STUDIES BY BRUCE C. HEEZEN AND MARIE THARP OF THE LAMONT-DOHERTY GEOLOGICAL OBSERVATORY

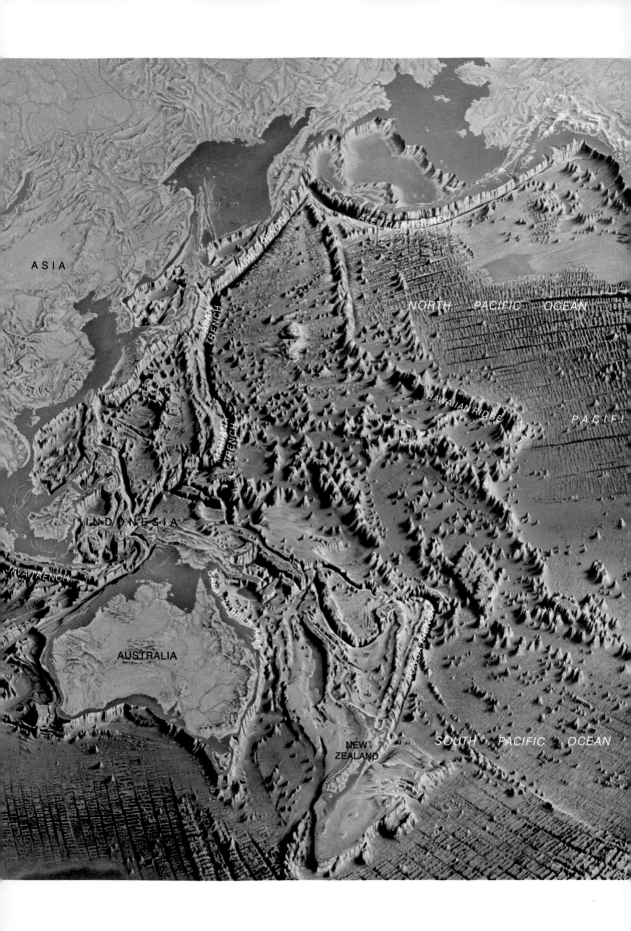

NORTH
AMERICA

MIDDLE AMERICA TRENCH

FRACTURE

ZONES

SOUTH
AMERICA

PERU-CHILE TRENCH

MID-OCEANIC RIDGE

skimmed above a cluster of gesturing lobsters and climbed full jet, telling me about a scorpionfish that lived up the hill.

"*Voilà! La scorpène.*" I saw nothing. "Right there," he pointed. "Don't you see him?" Again my land-oriented eyes failed to detect anything—the weedy coloration of the scorpionfish blended with its background. But Falco swung the saucer to within a foot of it, and I caught a glimpse of an eye and the outline of the mouth just before the fish stirred and whipped away.

Falco settled the saucer on a muddy plain covered with a tangle of rubber air hoses and power and communication cables. Something thumped on our steel hull, but Falco volunteered no explanation for the noise. Quickly he spun us around, simultaneously extending the hydraulic claw the saucer keeps clenched under her bow except when gathering samples. So neatly done that it betrayed rehearsal, an oceanaut shook hands with the saucer.

Through his face mask, I recognized the mischievous brown eyes of André Laban, mission chief of Conshelf Three. A black rubber suit with yellow tapes on the seams sheathed him from head to foot. His breathing hoses curled away into the murk. André, like the other oceanauts, inhaled "heliox," a mixture of helium and oxygen pumped to them from the undersea house when they worked outside. Without this special pressurized gas mixture, the menfish could not survive for long. At such a

Covering almost one-third of the globe, the Pacific could swallow the seven continents. The Mariana Trench reaches earth's greatest known depth, nearly seven miles below the surface of the sea. If placed beside Mount Everest, Hawaii's highest peak, rising more than 33,000 feet from the sea floor, would spire nearly a mile above the Himalayan giant. Walls of the fracture zones climb as much as 10,500 feet. The Mid-Oceanic Ridge sweeps around the world; its wide rift valley leads some scientists to believe the earth is splitting as the crevasse grows larger—but at an infinitesimal rate.

PAINTING BY HEINRICH C. BERANN BASED ON BATHYMETRIC STUDIES BY BRUCE C. HEEZEN AND MARIE THARP OF THE LAMONT-DOHERTY GEOLOGICAL OBSERVATORY

Diving in the Bahamas, Jo Starck and a companion breathe a mixture of helium and oxygen from Electrolungs. Developed by her husband, marine biologist Walter A. Starck II, the devices purify and recycle exhaled air, thus emitting no bubbles and allowing dives of up to six hours in water as deep as 550 feet. Entering the submersible decompression chamber hanging nearby, a diver can ascend immediately to the support ship for decompression. Below, an aquanaut sawing into a wreck wears a cryogenic lung. It warms liquid oxygen and nitrogen from −318° F. into a breathable gas by piping it through coils. Designed by Floridian Jim Woodberry, it allows its user to stay down more than six hours.

JERRY GREENBERG (BELOW) AND WALTER A. STARCK II

depth, the air we breathe on the surface would become highly dangerous because of its nitrogen content.

Two more oceanauts swam into our lights, and beyond them I caught my first glimpse of Conshelf Three. Nineteen feet across, the great black-and-yellow-checkered sphere rested like a forgotten beach ball on a chassis holding 77 tons of ballast. The steel house contained two stories, the lower for diving, sleeping, and sanitation, the upper for dining, communications, and sorting the many samples gathered from the sea floor.

In a happy salute to our friends, Falco glided up and over the sphere and brushed between two cylindrical escape chambers

on the chassis, each kept under pressure to carry the oceanauts to the surface in an emergency. He climbed above the house and then dived toward it, pulling up as I closed my eyes for the crash. Once he put the bow down so hard that my face flattened painfully against the porthole.

Falco enjoyed this noodling around. I knew he rarely could joyride in the saucer. Almost always he transports a scientist or an engineer with fixed designs on a particular spot for observation. Then poor Falco acts as a chauffeur. Today he swooped and swerved like a stunt pilot.

As the bow shot up again, I saw an overhang of yellow steel plate directly above us and in it a shimmering round mirror of

water — the always-open hatch of the undersea station. Water never rises above the hatch, because the heliox pressure inside equals the water pressure outside. Falco had expertly slipped the saucer between the uprights supporting the sphere, putting us under the hatch with no more than two or three feet to spare.

He circled the house again. It was about breakfast time, Falco said, time for the oceanauts to be downing their usual beakers of *café au lait* and hunks of bread smeared with jam, chestnut puree, and chocolate. Such an occurrence would have been pure hallucination a quarter of a century before when I first became addicted to following man's struggle with the sea.

Pushing through a "sky-light" of ice, the U.S. nuclear submarine Skate *surfaces 300 miles from the North Pole. Comdr. James F. Calvert took this remarkable photograph on March 22, 1959.*

Diving saucer designed by French Navy Capt. Jacques-Yves Cousteau encounters Aqua-Lungers six fathoms deep in the Red Sea.

Nemo *gives two observers wide-angle views as deep as 600 feet. Ballast and power supply housing rides below the plastic sphere.*

Those 25 years—especially the last ten—witnessed more progress in underwater exploration than all the years preceding. For centuries, none save divers for pearls, coral, shellfish, and sponges had firsthand knowledge of the contents of Neptune's domain. Dependent upon their ability to hold their breath, they could go on inspection tours of only a minute or two and cover sharply limited areas.

The desire for trade and new lands motivated many of our ancestors to venture on harrowing surface voyages. But for centuries even the saltiest old salts trembled at the thought of terrors awaiting them should they be unfortunate enough to "go below." Only the steady advance of scientific discovery eventually routed their fears—among them, the belief that frightful monsters skulked in the deep.

IMAGINE, if you can, a beast a mile and a half in circumference and covered with seaweed. The Norwegians believed it existed, and passed down stories about it from one generation to the next. The kraken, as they called this biggest of all creatures, continued to seize and swallow prey until the early 19th century. Not until then did it die in the minds of its believers.

Other imagined monsters included lobsters huge enough to drag down sailors with their claws, and serpents that crushed whole ships in their coils. If seafarers managed to escape these horrors, they undoubtedly thought themselves more vulnerable to the songs of silken-haired sea nymphs with a single mission: to lure sailors to their death.

Hazards real and imagined failed to daunt a scattering of daring visionaries who persisted through the ages with experiments calculated to offer men safe passage into the sea. Instead of myth, these farsighted, fearless innovators bequeathed new knowledge to those who succeeded them.

Conshelf Three and similar stations prove that man can inhabit, not merely visit, earth's last real frontier. "We have heeded our own discoveries and learned the lessons they present," Cousteau said of Conshelf Three. "The greatest of these is that if man is to make an undersea creature of himself, he must do it wholeheartedly and without a backward glance. As with a newborn infant, the umbilical cord must be severed.

"Within a few years," Cousteau continued, "we will eliminate entirely ties to the world above. Then, for the first time, oceanauts will have true freedom of the deep."

Such freedom, I thought, will allow chemists to work in a sea-bottom laboratory. Biologists will study fabulous creatures as yet unseen. Botanists will stroll underwater gardens, and geologists will decipher ocean history. Philosophers, poets, and artists will find room here too.

"Time to go back," Falco said, and I faced the inevitable end of my extraordinary voyage. After dropping ballast, he muttered entries into the log, and we climbed aloft. In hundreds of surfacings, Falco has emerged neither so near the mother ship that he imperils his craft nor so far away that he is ever out of sight of her.

Faint green light appeared and grew golden as the sun sparkled in the ports once more. Outside, a crewman made fast a line to haul us into range of the tackle that a few minutes later lifted the saucer into its cradle. I extended my arm up through the hatch and cleared my other shoulder through. Before me spread the familiar shores of France, gleaming in the winy October sun, but mentally I lingered in the hidden realm of the depth men. Falco and I had returned from a morning in the future.

Undersea house and workshop, Continental Shelf Station Three rests 328 feet down in the Mediterranean. Here six oceanauts stayed three weeks, leaving the steel sphere daily to carry out their assigned tasks. Hovering near the upper hatch, a crew member checks for rust and damage to power and communication cables. At the surface Captain Cousteau, director of the $700,000 project, kept constant vigil by television monitor as the team boldly advanced man's ability to live and work in the sea.

2

Man Invades the Sea

BY JAMES DUGAN

Mysteries of the underwater world lure Alexander the Great, King of Macedonia. Legend says he entered the sea in a "barrel" or "cage" of glass in the fourth century B.C. *An unknown artist painted this miniature 2,000 years later in India.*

MORE THAN FOUR CENTURIES AGO two adventurous Greeks, each carrying a lighted candle, hoisted themselves into what looked like a huge upended vase and entered the depths of the River Tagus at Toledo, Spain. Taut ropes lowered their ungainly apparatus while lead weights kept it from tilting and pulled it toward the bottom. No one knows how long they stayed underwater, but an eyewitness wrote that they emerged unharmed—with the candles still burning.

Thousands of spectators crowded the riverbanks that memorable day in 1538; many must have thought uneasily of sorcery. But in fact they were watching an exhibition of a primitive diving bell, staged before Charles V of Spain, ruler of the Holy Roman Empire. The two Greeks, dry and comfortable, breathed air trapped inside their crude submersible.

Their descent, one of the earliest reported with any detail and accuracy, drew attention to the diving bell, already an ancient but little-known device, and its use in salvage work on sunken ships spread slowly throughout Europe. Man was taking a long step into that dark underwater world that had fascinated him for thousands of years.

The sea itself, almost literally, is in man's blood, for the fluid that courses our bodies is much like seawater in chemical composition. Yet for us the oceans are a hostile environment. And the dangers increase with depth because the weight of water exerts great pressure on anything in it.

Today we can build submarines that hold this pressure out. Their crewmen live in an atmosphere like the air we breathe every day. In the diving saucer, for example, Falco and I carried a bit of our surface world down with us. Undersea observation chambers like the bathysphere and the bathyscaph give men the same protection. So do the clumsy experimental suits of metal that amount to wraparound submarines. This kind of packaged safety is fairly new in our invasion of the sea.

Ancient diving bells, on the other hand, and now undersea stations like Conshelf Three represent victories man has won in adapting his body to underwater pressure.

29

Man probably took his earliest steps into the sea in search of food. And what courage those first dives must have required! Imagine the boldest of the seashore tribesmen wading past minnows that skittered in the shallows, learning to paddle, swimming beyond the surf, and staring down through wavering green light. Finally, taking a deep breath, they ventured into a new world—a world where strange creatures darted and slithered among the rocks, where fearful dangers might lie just out of sight in the murk, or in the crevices of a coral reef.

Eventually men learned to wrest wealth from beneath the waves. Divers played an important role in the ancient world. Greeks and Romans worked in a thriving sponge industry, and their harvest had uses that seem strange today. Dipped in honey, sponges pacified babies; soaked in water, they became soldiers' canteens.

War provided still another reason for going underwater. In the fifth century B.C. when King Xerxes futilely attempted to conquer Greece, part of his fleet lay at anchor off Mount Pelion. Suddenly a severe storm struck without warning. The celebrated diver Scyllias and his daughter Hydna, who had deserted the Persian king and taken up the Greek cause, plunged into the waves. They swiftly began cutting anchor ropes, and the storm played havoc with the enemy ships.

Breath-holding divers still forage for treasure in many parts of the world—Japan, Korea, North Africa, Polynesia. In the Persian Gulf, I have sketched pearl hunters making lung-bursting dives as though special gear had never been invented. Holding stones to speed them down, they vanish in the turquoise water, snatch up oysters from the bottom, then rise gasping beside the boat, dark shoulders gleaming among vivid ripples in the savage heat. Yet even the most skilled of such "lung divers" can stay underwater only 90 seconds to two minutes at best, and they seldom go below 140 feet.

From Aristotle we get our first report of how man put his ingenuity to work to take air to divers under the surface. More than 2,000 years ago he wrote: "... they can give respiration to divers by letting down a bucket, for this does not fill with water, but retains its air. Its lowering has to be done by force." By swimming up into the large barrel-like containers and gulping a deep breath of the trapped air near the top, the divers could stay down longer.

And tales persist that in the fourth century B.C. Alexander the Great descended in what might have been a diving bell similar to an upended bucket. Though fables cloud the feats of the Macedonian king, it is highly possible that his courage and curiosity led him down into the sea. Legend says he rode in a "barrel" or "cage" of glass

Divers armed with compressed-air guns explore the eerie fantasy world created by Jules Verne in his book Twenty Thousand Leagues Under the Sea, *published in 1870. Their suits reflect gear invented five years earlier in France. Crewmen inside Verne's* Nautilus *gaze at a huge octopus.*

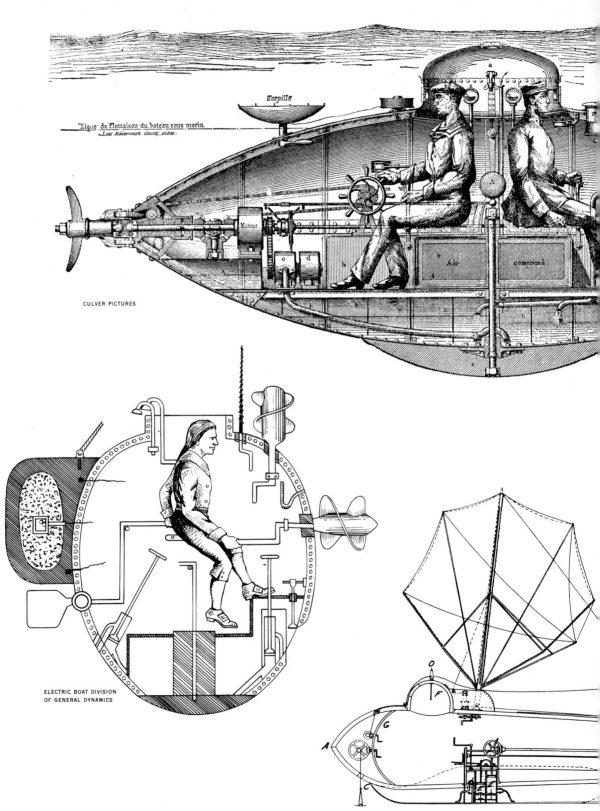

Torpille

Ligne de Flottaison du bateau sous marin.
Les Réservoirs étant vides.

Moteur

Air comprimé

CULVER PICTURES

ELECTRIC BOAT DIVISION
OF GENERAL DYNAMICS

FROM ORIGINAL PLANS OF ROBERT FULTON'S "NAUTILUS," 1798

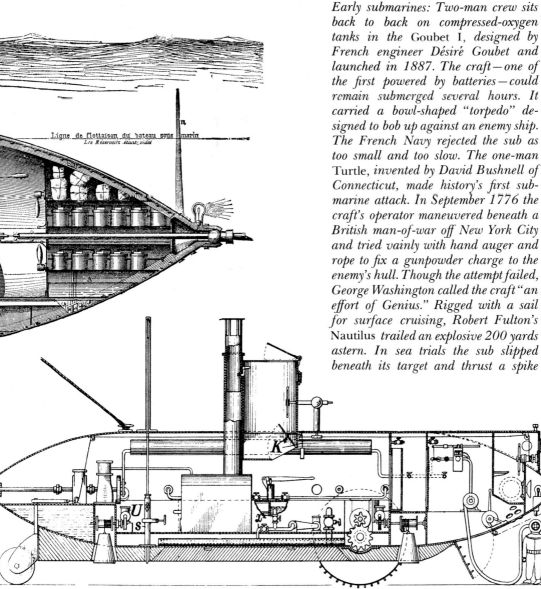

Ligne de flottaison du bateau sous marin
Les Réservoirs étant vides

ELECTRIC BOAT DIVISION OF GENERAL DYNAMICS

Early submarines: Two-man crew sits back to back on compressed-oxygen tanks in the Goubet I, *designed by French engineer Désiré Goubet and launched in 1887. The craft — one of the first powered by batteries — could remain submerged several hours. It carried a bowl-shaped "torpedo" designed to bob up against an enemy ship. The French Navy rejected the sub as too small and too slow. The one-man* Turtle, *invented by David Bushnell of Connecticut, made history's first submarine attack. In September 1776 the craft's operator maneuvered beneath a British man-of-war off New York City and tried vainly with hand auger and rope to fix a gunpowder charge to the enemy's hull. Though the attempt failed, George Washington called the craft "an effort of Genius." Rigged with a sail for surface cruising, Robert Fulton's* Nautilus *trailed an explosive 200 yards astern. In sea trials the sub slipped beneath its target and thrust a spike*

and ring into the planking. A cable, threaded through the ring, pulled the floating mine toward the vessel. Cross section of the ship's bottom (Q) shows the explosive (P) just before impact. France, Britain, and the United States spurned Fulton's underwater warship. Argonaut, *with an air lock for sending out divers, crawled on cast-iron wheels. Intended for undersea exploration, the boat made its first descent in Maryland's Patapsco River in 1897, operated by inventor Simon Lake. Air hoses for venting the gasoline engine sharply limited* Argonaut's *depth.*

Cranking the drive shaft, crewmen power the Confederate submarine Hunley *as the captain peers from the conning tower. The drawing exaggerates the size of the craft. Actually, she carried just nine men and had an inside height of only five feet. In an 1863 painting (top), soldiers guard the converted steam boiler in a Charleston, South Carolina, dry dock. The* Hunley *became the first submarine to sink a ship in war when she rammed a torpedo into the U.S.S.* Housatonic *in Charleston Harbor in 1864. But the explosion of the charge sent the sub to the bottom as well.*

—a material, remarkably enough, that may form the hulls of future deep submergence vehicles, because its molecular structure increasingly resists compression stresses as the weight of the water increases.

With a drinking glass, you can easily make your own "diving bell." Turn the tumbler upside down, hold it level, and push it into water. If you are careful not to tip the glass, it will trap all the air inside, forming, in effect, a model of Alexander's cage. If you could push the tumbler several feet down, the air inside would be compressed, reduced in volume by the pressure of the water in the mouth of the glass.

In the ocean, a quart of air trapped at the surface will compress to a pint at 33 feet, half a pint at 100 feet—whether in an open bottle or in the human body. And this compressed air can be deadly to a diver. My friend Capt. George F. Bond, USN, a specialist in underwater medicine, once startled me by pointing out how easily it could have killed one of the men I most admire: English astronomer Edmond Halley, who explored both stars and sea.

In 1691 Halley patented a diving bell made of wood and coated with lead to make it sink. With no ill effects from the "condensed air," as he called it, Halley and four companions spent more than an hour and a half in the bell at about 60 feet.

"But if Halley had stayed at that depth just a few minutes longer," George explained, "he might have died when he came up. If such a famous man had been

lost in that kind of accident, the practice of diving could have suffered a severe setback."

No one understood then that a person breathing compressed air for an extended time below 39 feet risks paralysis and death unless "decompressed" before returning to the outside air. Early bell and helmet men usually worked short shifts in shallow water, and no doubt many of them escaped trouble by sheer good fortune.

Designs for diving helmets appeared

centuries ago. About 1500 Leonardo da Vinci sketched a helmeted figure breathing through a tube buoyed by a float at the surface. Anyone wearing such gear would have great difficulty—a man just four feet down could not draw air through the tube because his lungs could not overcome the external water pressure. But if compressed air could have been pumped into the helmet, it would have equalized the pressure and allowed the diver to breathe easily.

Credit for a reliable system of supplying compressed air to divers goes to Augustus Siebe, a German engineer who found England "an inventor's paradise" and settled there. He developed a "closed suit" made watertight by rubber cuffs and a collar. Divers liked his new apparatus because they could wear clothing under it and keep out the cold of the sea. The suit soon became standard, and today it serves—with some modifications—as effectively as in 1837.

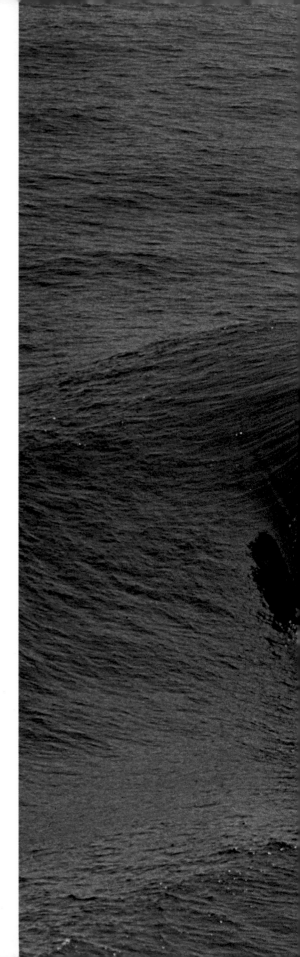

The same year one of the first diving-equipment firms in the United States, the presently named Morse Company of Boston, went into business. It's still going strong.

I visited the company's eighth-floor workshops on Sleeper Street, hard by Boston Harbor, and found a graying coppersmith making helmets with breastplates much as his fellow craftsmen had done a century before. I watched him hammer a copper disk around a steel form until it took shape as a breastplate, then solder on a neck ring for fastening it to the helmet. He fitted the helmet itself with a glass faceplate and three other windows, an air escape valve, and two goosenecks. The first would hold an air hose, the second a telephone line.

"We build them to last," said William P. Dugan, president of Morse's, as he smiled at my surprise over the unvarying art of helmet-making. "The other day a man brought in an old helmet for repair. We checked our files and found that we had sold it to his grandfather. Three generations have used it."

While the helmet-and-hose divers of the mid-1800's plodded about in the silt of harbors and riverbeds, compressed air found other uses. Workmen in caissons breathed it as they dug tunnels or sank bridge piers. Returning to the surface, both groups often showed the same symptoms: rashes and itching, choking, and agonizing pains in joints and muscles. Their bodies twisted into the dreaded bends that no doctor could relieve until a French physiologist, Paul Bert, discovered the cause.

An energetic and unorthodox man, Bert first investigated the effects of rarefied air on mountain climbers and balloonists. Then he turned his attention to other extremes of pressure, those of the world beneath the sea.

Roiling water cascades from diving planes on the fin of the nuclear submarine Shark *as it cruises in the Atlantic. Another atomic sub,* Triton, *circled the globe in 1960. Without surfacing, it logged 30,752 miles in 61 days.*
BURKE UZZLE

In 1878 he traced the torment of bends to its source—nitrogen breathed under pressure. This heavy inert gas makes up four-fifths of our atmosphere; as a rule it slips harmlessly in and out of our lungs. But a diver 33 feet down breathes twice as much nitrogen, at 66 feet three times as much. Pressure drives it into his blood and tissues. When a diver has absorbed all the gas his body will hold, he is "saturated."

As long as he remains under enough pressure—at saturation depth—the dissolved nitrogen causes no difficulty. If he comes up before too much of it accumulates, as Halley did, his lungs can get rid of it. But let him ascend too far too fast and the gas will cause his blood to start frothing like a well-shaken bottle of soda pop. The explosion of bubbles will clog the blood vessels, torturing, perhaps killing him.

PAUL BERT BELIEVED that a diver could reach the surface in safety if hauled up slowly enough for his lungs to eliminate the excess gas from his body. From 100 feet, for example, an hour-long ascent would ward off the bends.

Divers learned they could relieve a victim of the malady by returning him to the water and hauling him up slowly. Or they could place the patient in an airtight chamber, put him back under an air pressure equal to his diving depth, and gradually reduce the pressure.

I have never seen anyone afflicted with the bends, but in Monaco my wife, Ruth, and I learned that an acquaintance, a young biologist, had survived the terrible ordeal. After five days' decompression treatment, the hospital had just released him. We couldn't get over how pinched and shrunken he looked.

"He actually seems physically smaller," Ruth remarked.

This phenomenon of saturation, which can prove so deadly to divers, led to the startling discovery that men can adapt themselves to pressure at a given depth and *live* underwater. All a diver needs is a sea-floor dwelling that he can enter and leave at will, with an inside atmosphere compressed to equal the pressure of the surrounding water. George Bond suggested this concept of "saturation diving" in 1957, and men have since lived and worked in the sea for a month without surfacing.

The search for diving safety has long occupied scientists. In 1906 Professor John Scott Haldane of Scotland employed helmet-hose suits and refined Bert's discoveries into systematic decompression tables. Many years later I had the good fortune of meeting some colleagues of Haldane in his historic research with the British Admiralty Deep Diving Committee. One, Guybon C. C. Damant, became a good friend. Much as he disliked publicity about his remarkable dives, he obligingly told me of his work as a young gunnery officer in the Royal Navy.

"Unlike most of my class of budding gunnery lieutenants," Damant said, "I found going underwater to be a delightful experience and infinitely preferred it to the study of ballistics and gun drill."

Haldane declared that divers could ascend faster than Bert had recommended, and, in fact, *should*—in stages. For instance, a diver 100 feet down would be hauled quickly to 50 feet and, after a stop there, rise to 25 feet for another stage of decompression, halving the pressure at each level. This way he would eliminate nitrogen faster and reach the surface sooner.

To test the theory, Damant and Gunner Andrew Y. Catto got into heavy helmet dress and dived from H.M.S. *Spanker* off the southwest coast of Scotland. The first day they descended to 138 feet, and later as deep as 180 feet. Each time they came up in stages—exhausted but without having the bends.

Finally, Damant and Catto volunteered

Diving bell patented in 1691 by English astronomer Edmond Halley enables salvors to recover sunken cannon. Weighted barrels take air down to the lead-coated wooden chamber. Halley also devised a technique of supplying air to helmet-hose divers, conveying it from the diving bell "in a continued stream by small, flexible pipes."

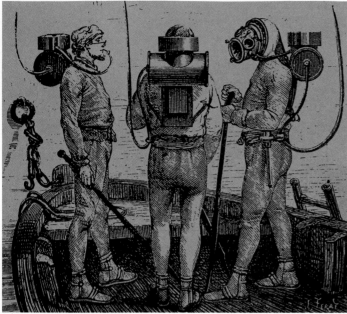

"Diving Armour," patented in America in 1830, resembles an oversized snorkel. No record exists that anyone ever built or tested the device.

Hoses and canisters supply air to divers preparing to submerge with Aérophores *developed in France in 1865 by mining engineer Benoît Rouquayrol and Navy Lt. Auguste Denayrouze. The reservoirs, filled with pressurized air, would give a diver several minutes to save himself*

if his air line parted. In 1680 Italian physicist Giovanni Borelli envisioned gear (below) that would free man from the surface, "purifying" its own air supply by passing it through a tube cooled by sea water. His concept included a hand-cranked ballast tank and clawlike foot fins.

to go below 200 feet, deeper than man had ever ventured before. "Mark you," Damant told me, "we were not attempting to set any records. We were simply trying to provide a greater working range for compressed-air diving."

For the unprecedented plunge, *Spanker* moved into deeper water, in Loch Striven. There Damant, then Catto, planted their lead boots on the bottom, 210 feet down. Their decompression stops completed, they surfaced without harm from a depth no other man would reach for nearly a decade.

The Committee's tables still form the basis of all compressed-air diving guides to prevent the bends. But the diver faces still another problem, the insidious danger of nitrogen narcosis—"rapture of the deep."

Nitrogen can intoxicate a diver almost like alcohol, and the effect increases with depth. I have often asked divers how it

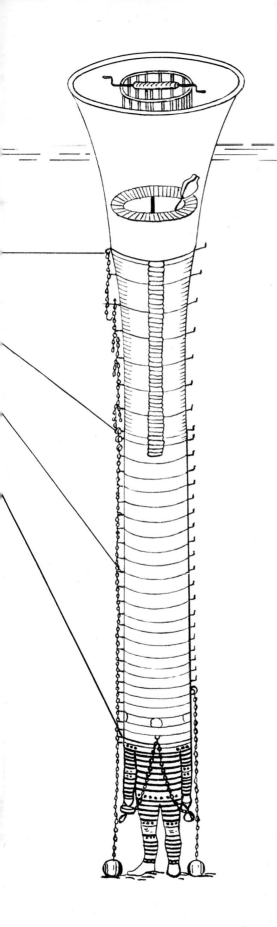

ELECTRIC BOAT DIVISION OF GENERAL DYNAMICS (LEFT);
MARINER'S MUSEUM (ABOVE); AND CULVER PICTURES

Hoping to salvage the Lusitania, *American diver Benjamin Leavitt (above) publicized a semi-armored suit in 1922. His plans failed for lack of funds. The boiler-shaped rig (below), built in Germany in 1797, allowed a diver to stay briefly in shallow water.*

feels. Their answers jibe with Captain Cousteau's: "The first stage is a mild anesthesia, after which the diver becomes a god. If a passing fish seems to require air, the crazed diver may tear out his air pipe or mouth grip as a sublime gift."

Difficulties with compressed air led to the testing of other breathable mixtures. One is the helium-oxygen that oceanauts used at Conshelf Three, and another is hydrogen-oxygen. But these combinations carry hazards of their own.

If too much oxygen is added to hydrogen, the mixture becomes highly explosive. Helium averts narcosis but causes rapid loss of body heat in cold water and distorts the shivering diver's speech into a high-pitched garble. In spite of these drawbacks, however, helium seems the best choice.

Men underwater have long relied on simple signals to the surface, tugging at hose or rope by prearranged code. The invention of the telephone provided some

with a more sophisticated communication system. But lines can snag on rocks and wreckage and trap the man they protect. Only by breaking these tethers could divers move about with ease.

Attempts to free men from all dependence on the surface began centuries ago. Leonardo sketched a design for an underwater costume with a "breastplate of armour," sacks of sand for ballast, and a "wine-skin to contain the breath"—the first serious effort to devise self-contained underwater breathing equipment.

Almost 200 years later, another Italian, Giovanni Borelli, tried to extend the diver's range with more complex gear. He believed exhaled air could be purified and breathed again if simply channeled through a copper tube cooled by sea water so condensation in the tube would trap "impurities."

His insight outran the technology of his day. For years specialists rejected his concept as one that couldn't possibly work. But

Silhouetted by filtering sunlight, a breath-holding snorkel diver stalks fish near a Mediterranean reef. Carrying no air supply, he can make only shallow, brief excursions with face mask and spear. A helmeted diver, who can stay down for hours, services a gas well 240 feet down off California.

George Bond once told me: "This principle of cold-purification makes good sense. You know, Cousteau experimented with it by freezing out air impurities in Conshelf Three, and eventually it will be used in all undersea habitats."

Around 1831, a Brooklyn machinist named Charles Condert dreamed up a self-contained outfit for himself—and it did work. To store compressed air he employed a six-inch-diameter copper pipe closed at the ends and bent to circle his waist. Air reached his lungs through a tube leading into his closed dress. He often clambered alone into the tricky currents of the East River, and walked about while excess air bubbled from a small hole in the top of his suit. One day a few large bubbles died unnoticed at the surface—the tube had broken and Condert had drowned. Brief notices of his work tantalized readers on both sides of the Atlantic.

As a boy I read and reread *Twenty Thousand Leagues Under the Sea*, a book that made an ingenious diving rig world famous. Jules Verne invented heroes for his submarine *Nautilus*, but not their basic gear. A French mining engineer and a naval officer had developed that five years earlier, in 1865.

Benoît Rouquayrol and Lt. Auguste Denayrouze equipped their helmet-hose suit with a double-chambered air tank. Compressed air pumped from the surface filled the reservoir on the diver's back, and he inhaled through a hose and mouthpiece. With this new gear, he could breathe more comfortably than ever before, for the inventors had equipped their suit with a demand regulator valve, their most important contribution. It fed air as needed from the reservoir, at exactly the right pressure.

Meanwhile, a wiry and prankish Englishman named Henry Fleuss was taking more chances than he knew. In 1868 he patented

Propelled by flippered feet, a free-moving swimmer 90 feet down in the Aegean Sea glides past a Greek hard-hat sponge diver encumbered with air hose, lifeline, and weights.

BOB KENDALL

45

a fully self-contained apparatus that supplied pure compressed oxygen. He would stay down more than an hour, inhaling the same oxygen again and again while a chamber of caustic potash absorbed carbon dioxide from the breath he exhaled. What Fleuss did not know was that below 25 feet or so pure oxygen becomes so toxic it can cause convulsions and death.

Ten years later Paul Bert published a warning about this poisonous effect. "Pressure," Bert explained, "acts on living beings . . . as a chemical agent changing the proportions of oxygen contained in the blood, and causing . . . toxic symptoms when there is too much."

Luckily, Fleuss escaped harm because he dived in shallow water. Equipment like his demanded the utmost caution, but war created a use for it. Frogmen in World War II wore oxygen rebreathers patterned after Fleuss's closed-circuit system. The gear leaves no telltale stream of bubbles, allowing the underwater combatant to move undetected in enemy waters.

Warfare also brought key developments in the submarine. Beginning with a one-man hand-propelled craft in the American

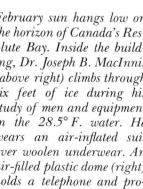

ANDRES PRUNA (BELOW) AND JAMES A. SUGAR

February sun hangs low on the horizon of Canada's Resolute Bay. Inside the building, Dr. Joseph B. MacInnis (above right) climbs through six feet of ice during his study of men and equipment in the 28.5° F. water. He wears an air-inflated suit over woolen underwear. An air-filled plastic dome (right) holds a telephone and provides an emergency refuge.

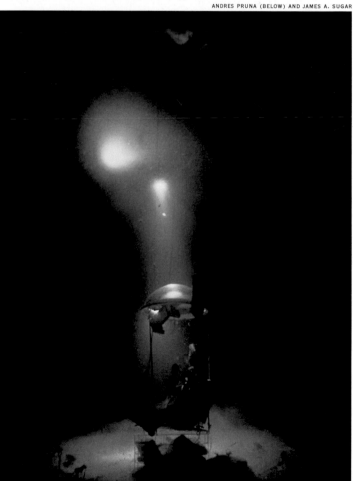

Revolution, underwater boats have evolved through the Civil War and two world wars into the nuclear fleets of today. Now a variety of non-military submersibles are playing their part in the struggle of scientists and engineers to win freedom of the depths for mankind.

For the diver, freedom means more time underwater and greater mobility. The breath-holding diver can swim about easily, but only for seconds; the helmet-hose man can stay down for several hours, but only in a restricted area.

IN THE 1920's AND 1930's, French Navy Comdr. Yves Le Prieur devised a system that transformed the diver from a weighted plodder at the bottom into a free swimmer. He supplied a compressed-air tank and hand-operated breathing valve that, combined with rubber foot fins and a light face mask, allowed the underwater man to move almost effortlessly in any direction.

Now one important step remained—the development of a reliable demand regulator valve simpler and more compact than that invented by Rouquayrol and Denayrouze. This would free the diver's hands of the task of valving air from his tank.

At last, in 1943, the "demand valve" ushered in a new age of underwater exploration. Cousteau, then a lieutenant commander in the French Navy, and Emile Gagnan, a Parisian engineer, perfected the device. Their success put safe diving for sport within the reach of virtually everyone.

I first saw their gear, the Aqua-Lung, in the motion picture *Épaves (Sunken Ships)* in 1944. The film showed Cousteau and his underwater companions, Frédéric Dumas and Philippe Tailliez, swimming fishlike among weed-encrusted wrecks with large compressed-air cylinders on their backs. I watched it twice. The experience so fascinated me that I determined to seek out the man behind this marvelous invention.

I met him in London, and wrote an article on his adventures. Though I sent the story to dozens of publications, they showed no interest. Finally, three years later in 1948, a new magazine, *Science Illustrated*, pub-lished it, and soon I was opening a lot of mail requesting information about the Aqua-Lung. The article caught the attention of Lt. Comdr. F. Douglas Fane, head of an underwater demolition team, and work he later performed with the new apparatus won him a Navy commendation.

The Aqua-Lung permitted men to move weightlessly in the sea, with a range never possible before. Cousteau sums it up well: "From this day forward we would swim across miles of country no man had known, free and level, with our flesh feeling what the fish scales know."

Perhaps someday men will roam the depths even more freely, with their lungs breathing water as the gills of fishes do.

. At Duke University, Dr. Johannes A. Kylstra, who originated the concept, is continuing experiments begun in 1961. In the laboratory he determined that dogs can live in a saline fluid up to one hour and emerge safely. [*Editor's note:* In 1968 Dr. Kylstra successfully flooded the lungs of diver Francis J. Falejczyk with water about as salty as human blood. Conscious throughout the experience, Falejczyk reported no unpleasant sensations. For their work, proving that such a saline solution will not damage lung tissue, the two received the Marine Technology Society's 1970 Lockheed Award.]

Says Captain Bond: "A diver breathing saline water may someday be able to descend more than two miles and pop back up without needing decompression."

No one can predict what techniques man may devise to explore the depths—or what wonders he will find. Surely the majesty of the ocean will continue to inspire him: "this great and wide sea, wherein are things creeping innumerable, both great and small beasts," just as the Psalmist has said.

Sea Scooter, its twin screws driven by battery-powered electric motors, tows its developer Carl Gage above a coral garden off Key Largo, Florida. The 125-pound aluminum scooter can take a diver silently to depths of 100 feet and pull him for three hours at speeds up to three knots.

3

Revealing the Ocean's Secrets

BY ROBERT C. COWEN

Instrument booms project from FLIP—
*for floating instrument platform — as
oceanographers collect data off southern
California. Its stern flooded with 600
tons of seawater, the unusual laboratory
tips vertically for stability; the slender
hull protrudes 300 feet into the sea.*

AN AIR OF ANTICIPATION stirred our small gathering of science writers. Professor J. Tuzo Wilson of the University of Toronto, an internationally prominent earth scientist, turned toward a pot that simmered over a gas flame. I craned my neck for a better view, and caught the aroma of tomato soup!

"The science of the earth," he said, "is in a state of intense excitement." He tipped the pot as he spoke, and I saw what I have often observed in my own kitchen — a light froth floating on top of a mass of liquid that roiled and overturned. Soup heated near the bottom of the pot rose to the surface, spread, cooled, and sank again to complete the motion physicists call convection.

"We have been accustomed to regarding the earth as a solid object, like a stone," Dr. Wilson continued. "It is hard to visualize, but actually it may bear much more resemblance to a pot of soup convecting on a stove. Such a pot has a two-phase surface of liquid and froth. The floors of the ocean, like the soup, are forever turning over and being renewed while the continents, like the froth, persist and grow."

Older theories presented our planet as a rigid, almost static body jiggled occasionally by slight disturbances, holding its oceans in basins of solid rock. Now geophysicists think of the earth's crust as ruptured into huge, constantly shifting plates: the theory of plate tectonics. Derived from the same root as architect, tectonics refers to the art or science of building; plate tectonics deals with the interaction of plates in forming and re-forming the geological structure of the earth.

Two plates may bump together and buckle to form a mountain range. Elsewhere one plate may slip under another, and ride downward into the earth's hot, plastic mantle; in this instance, earthquakes may result. Part of the Pacific floor slides under Japan and the northwest Pacific Ocean at a rate of about four inches a year — an extremely rapid rate, geologically speaking.

What pushes the plates around? Scientists debate this at length, although there is

51

Moored to a breaker-lashed cliff, the British corvette Challenger *sends boats along a lifeline to explore St. Paul's Rocks in the mid-Atlantic. Trailblazer in oceanography, the ship sailed from Portsmouth, England, on December 21, 1872, for a voyage that spanned three and a half years and 68,890 nautical miles. Fitted with an auxiliary steam engine in addition to her sails, she carried a crew of about 240 and cruised with 16 of her 18 guns removed to make room for equipment to take soundings and bottom samples. Expedition scientists plumbed the great ocean basins, recorded sea currents and temperatures, and gleaned from the deep 715 new genera and 4,417 species of plants and animals. A twine dredge bag (left), dragged along the ocean floor, netted many of the specimens. Trailing behind the bag, eight hempen tangles swept up bits of coral, sponges, and other organisms.*

Chief scientist of the Challenger expedition, Scottish naturalist Charles Wyville Thomson headed a six-man ocean-research team. In the "natural history work-room" (below) on the ship's main deck, the staff examined its findings and preserved samples in test tubes and bottles.

"THE VOYAGE OF THE CHALLENGER, THE ATLANTIC," VOLUME 1, 1877

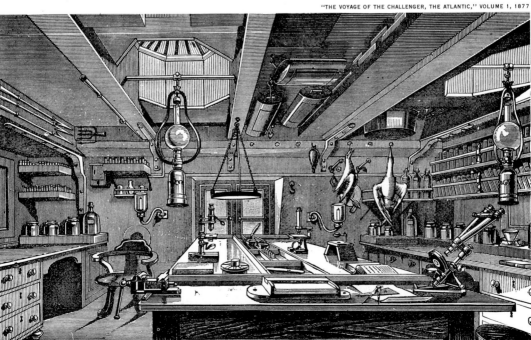

general agreement that slow-flowing currents of molten rock deep in the earth are responsible. Specialists explain the movement variously, in terms of pressure from the Mid-Oceanic Ridge; a shifting of plates away from hot spots; a tug from the sinking portion of an oceanic trench; or general convection currents within the mantle.

Experts are refining their concepts of the ocean waters, too. For oceanography is the scientific study of the sea in all its aspects. Once when I pressed Dr. Roger Revelle, formerly director of the Scripps Institution of Oceanography at La Jolla, California, to

THE AUTHOR: *Robert C. Cowen, feature editor of the* Christian Science Monitor, *received a Master of Science degree in meteorology from the Massachusetts Institute of Technology in 1950. He is the author of* Frontiers of the Sea, *a comprehensive study of the science of oceanography, first published in 1960.*

define oceanographers, he smiled wryly and said, "They're sailors who use big words." These specialists include biologists, chemists, engineers, geologists, and physicists. They must move deftly between the tranquillity of their laboratories and the rigors of life at sea—not the least of which, for many a landlubber, is seasickness.

When H.M.S. *Challenger* sailed from England in 1872, naturalists aboard had to cope with both seasickness and naval etiquette. They returned 3½ years and 68,890 nautical miles later, completing the first great voyage of deep-sea exploration.

Their chief, Charles Wyville Thomson, and other men of science had persuaded the government to sponsor the expedition. The Lords of the Admiralty lent the corvette *Challenger*, which was then specially fitted for research.

Oceanographers still enjoy stories of that

mission: At first even the cabin boys gathered eagerly when a dredge broke the surface, to watch for rarities or even "living fossils." But as H. N. Moseley, the coral expert, noted, "the same tedious animals kept appearing from the depths," and at last even most of the scientists went on eating when the dredge came up during dinner.

At 362 observing stations, officers and crew helped the six civilians record weather, current speeds, and water temperatures. Wherever they could, they dragged a trawl. Thomson's enthusiasm never flagged; he patiently examined every cuttlefish from every haul. The men spent hours lowering a weighted rope to take soundings, and their parrot Robert learned to chant, "What! 2,000 fathoms and no bottom!"

Reports of their work fill 50 large and useful illustrated volumes with details on seawater, marine life, and submarine geology. But from isolated soundings with lead and line, no one could glimpse the sea floor's true complexity.

Oceanographers used to picture the land of the abyss as a smooth, monotonous plain, shrouded by sediments that covered any original features.

The pace of research has quickened since high-pitched sound pulses replaced the miles of rope and wire—unwieldy and subject to breaking—that early oceanographers payed out to measure the depths. Now a pinging echo sounder logs a continuous profile of the sea floor.

NATIONAL GEOGRAPHIC PHOTOGRAPHER BATES LITTLEHALES

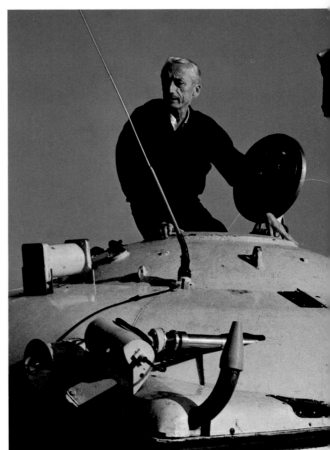

THOMAS J. ABERCROMBIE, NATIONAL GEOGRAPHIC STAFF

Captain Cousteau emerges from his diving saucer after a cruise hundreds of feet down in the Caribbean. Battery-powered hydrojet nozzles, one on each side, propel the vehicle. The research ship Calypso *(left) serves as floating headquarters for the undersea pioneer's worldwide expeditions. A stern winch holds the saucer.*

Frothy wake streams from a "monster buoy" en route under tow to a moorage 1,500 miles north of Honolulu. From the top of its 50-foot-high mast to 1,500 feet down on its mooring line, the instruments of Ocean Data Station Bravo re-cord oceanographic measurements and radio the data to computers ashore. The National Oceanic and Atmospheric Administration hopes to create a network of such buoys to aid in oceanographic research and weather forecasting.

Brass Nansen bottle will trap a water sample in an Indian Ocean deep, and record its temperature.

Ocean bottom seismograph, its protective steel sphere removed, records signals from undersea detonations.

On a platform of the U. S. Navy's oceanographic research ship Sands, *scientists lower a weighted tube to extract a core sample of sea-floor sediment from the Atlantic. Wearing headphones for intership communication, oceanographer Larry Hawkins (left) watches as an oscilloscope monitors the position of a deep-sea camera.*

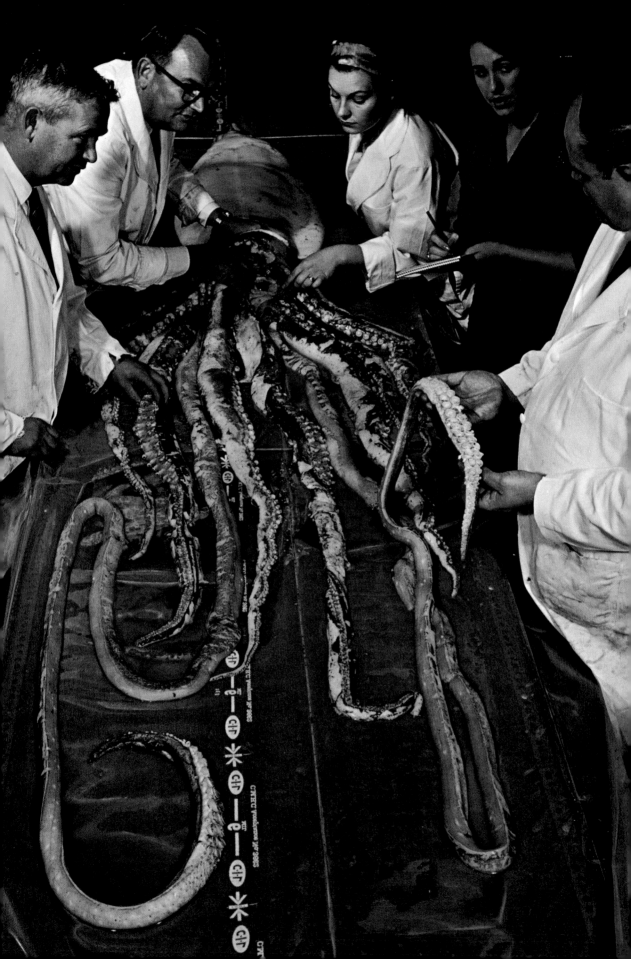

Also, thanks to the sophisticated seismic profiler, geologists can analyze the land far beneath their ships from the patterns of returning sound waves. Signals from an electric spark or the rumbling shock waves from an explosive charge enter the sea bed, where layers of different densities reflect them back to the surface at different time intervals.

An even more advanced device, the side-looking sonar, promises detailed mapping of the sea floor. Like aerial cameras, such sonars scan a wide swath. Suspended near the bottom of the ocean, they show details measuring only a few feet across while mapping a strip many hundreds of feet wide. Thus, gradually, a more dramatic seascape has appeared.

Its mountains include earth's longest range—the last to reach the maps. "In 1954," Dr. Bruce C. Heezen of Columbia University's Lamont-Doherty Geological Observatory recalls, "I sat down with Marie Tharp and Maurice Ewing, who led the Mid-Atlantic Ridge Expeditions sponsored by Columbia, Woods Hole Oceanographic Institution, and the National Geographic Society back in the 1940's. We pored over thousands of worldwide soundings collected by our research ketch *Atlantis*. We suspected that the Mid-Atlantic Ridge might link together with other ranges to form a titanic feature, so immense that no one had imagined it."

I wrote enthusiastically about their theories, and got friendly warnings that the "Lamont crowd" was jumping to conclusions. Today their conclusion ranks as one of the major ocean discoveries of our time.

Many scientists believe this ridge system marks a zone where sea-bed crustal plates move apart. Temperature studies made along the ridge offer further evidence that molten material welling up from the underlying mantle forms new sea bed here. Deep ocean trenches border some plates: Where the Pacific sea floor plate off South America underrides the continental plate, the Peru-Chile Trench makes a deep gash almost 2,000 miles long.

As they shift, plates form mountain

Marine biologists at work: A diver lifts a rare live chambered nautilus from a trap off New Caledonia. In Newfoundland (opposite) doctors dissect another rare find, a 21½-foot-long young giant squid. Below, sensitive instruments measure the metabolism of bottom animals.

ranges, open up new oceans, close old ones, and carry continents about. The North Atlantic probably opened in this way over the last 200 million years. Conversely, the Himalayas probably rose when the north-ward drifting plate carrying India slid under Tibet and closed an ancient sea.

In 1912, German meteorologist Alfred Wegener aroused interest in the notion of drifting continents, attempting to prove that the break-up of the ancient super-continent Panagea formed our present continents. Anyone can see how neatly South America's bulge would snuggle into West Africa's bight. But no one could explain what might move such huge land masses until the concept of the dynamic earth began to take shape. Today most geophysicists accept Dr. Wilson's sugges-

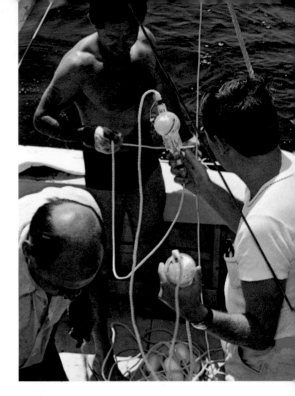

tion of shifting crustal plates, as supporting evidence of continental drift piles up.

Gathering data is a slow process. And most oceanographers still work from the deck of a surface ship. Dredges bring up rocks, animals, and plants; instruments of steel plunge into bottom ooze to measure the heat flow from the earth's interior, and hollow tubes punch out long sediment cores.

These cores—averaging 30 feet in length

Aboard Calanus, *research vessel of the University of Miami's Institute of Marine Science, inventor Shale Niskin (in trunks) hauls in recorders that measure ocean-current fields. Below, the instruments record bottom currents in the Straits of Florida. A hard-candy disk links bulb and conical weight. When the candy melts, the pink-topped tubes will pop to the surface.*

—stretch backward into time and give a glimpse of the ancient world. Settling over the eons, the remains of plants and animals mingled with meteorites and other particles to build up the sea bottom.

Examining pinches of sediment under the microscope, I have seen perfectly preserved shells of tiny sea creatures long extinct—radiolarians as lacy as snowflakes, foraminifers that lived only in cold water and others that flourished in warmer seas.

Paleontologists have identified these animals from their characteristic shapes, and, by distinguishing between species that lived in warm or cool water, they can trace variations of ocean temperatures in the past.

"At Lamont," geologist David B. Ericson told me, "we have one of the world's largest deep-sea core libraries. By comparing

Scientists study pink shrimp and deep-sea sediment at the Institute of Marine Science. Dr. David A. Hughes (below) watches the response of shrimp to current and salinity changes for clues to migration patterns. A National Geographic Society grant supported the project. Marine geologist Walter E. Charm examines sediment cores refrigerated to prevent mold.

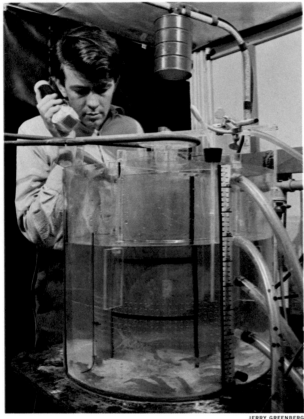

JERRY GREENBERG

strata of sediment, we have traced the advances and retreats of the glaciers for the last 1½ million years."

To geologists, who think in terms of hundreds of millions of years, this seems like yesterday. Working from the *Glomar Challenger,* a specially designed drilling ship, researchers have probed many areas of the sea floor. Like the men aboard the H.M.S. *Challenger,* who studied the ocean a hundred years earlier, *Glomar Challenger's* scientists are discovering bits of information that revolutionize our thinking about the oceans.

As part of the Deep Sea Drilling Proj-

ect, *Glomar Challenger's* drill has bored more than 3,000 feet into the sea floor in water 20,000 feet deep—requiring a vertical "drill string" of pipe more than four miles in length! Often it passes through the entire sediment burden into the underlying rock.

Nowhere do sediments seem older than about 160 million years, a fraction of continental ages. To scientists, this confirms that the ocean floor has been renewed throughout geological time.

Sometimes the sediment record seems frustratingly garbled. Dr. Anthony S. Laughton of Britain's National Institute of Oceanography showed me data from one sample that had appeared to be geological gibberish. "We had expected an orderly layering," he explained. "Instead we found a jumble. Ancient currents destroyed the sediment structure. But they

left valuable records of their own. We have learned to read this sample backward in time to decipher how North Atlantic waters circulated in past geological epochs."

While fascinated with such history, Dr. Laughton's colleague Dr. Nicholas C. Flemming focuses on the geological "now." He spread out a map of ancient Mediterranean settlements to show how he constructs a "motion picture" of the long-term rise and fall of the earth's crust. Ruins of ancient seaside habitations, some of which now stand high above the water line or lie beneath the waves, give him reference points: He can interpret their changing relationship to sea level in terms of actual rise or subsidence of the land.

The eastern part, from about longitude 18° to 20° east, reflects the turmoil of one of the world's most active seismic regions. He finds the western Mediterranean rela-

tively stable, and detects the vertical crumbling where crustal plates grind together. Southwestern Turkey is subsiding; Rhodes is tilting into the Aegean; and western Crete is tilting as a rigid block, up at the southwest and down into the Aegean.

In spite of the help of modern probes and drills, the oceanographer often must grope for what lies below the ocean floor. What he cannot see, he has learned to hear with instruments that interpret for him.

Until recently, scientists who study ocean circulation lacked such tools to give a direct view of their subject. They sought their clues in the waves and currents. Stirred by the winds and by the tide-ruling forces of the sun and moon, the sea presents a changing panorama—sometimes regular, sometimes chaotic—with rhythms that often signal meanings all their own.

Booming breakers on our western shores

63

Opening like a book, a water sampler helps contribute a chapter to man's knowledge of the sea. Triggered at a predetermined depth, the metal covers pop open and a sterile plastic bag traps water, then seals automatically. In the Straits of Florida, a marine biologist winches the apparatus aboard Calanus *(right). At the Institute of Marine Science, research aide Rose Cefalu studies microbe cultures produced from water samples brought up from varying depths.*

hint of typhoons that raged thousands of miles away. Choppy whitecaps in a channel may tell of ebb tide flowing against opposing winds or swell. Incoming waves along a beach may reflect the contour of the offshore bottom—they rise higher over banks and sandbars, or break with lower crests after crossing a submarine canyon.

Everywhere, sun and moon pull the sea into motion, but the shape of basins and shores determines the local timing and range of the tides. For instance, in the Gulf of Mexico the tide rises and falls once a day, and only by a foot or two; at Río Gallegos, Argentina, it changes twice a day, by as much as 45 feet 7 inches.

Quite apart from tides, however, the interlocking water masses of the oceans are forever on the move. Sometimes they rush

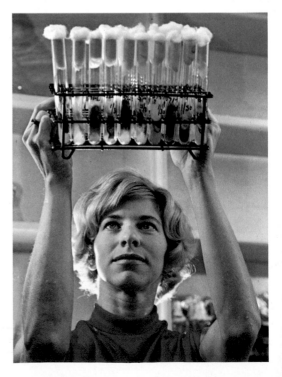

along with enormous power, as when the Gulf Stream pours past Miami and flows up the coast. Sometimes they barely creep, like the cold bottom currents that may take decades to complete their journey. But somehow every drop of water will find its way to almost every part of the sea's domain.

This movement of the waters stirs the oceans continually. It regulates the weather. It creates upwelling currents that bring nitrates, phosphates, and other minerals from the depths, fertilizing the plants that feed the world's great fisheries.

"To build a comprehensive science of the sea, we must first understand its basic circulation"—all oceanographers say this. For centuries shipmasters have logged currents, yet scientists have scarcely begun to sort out all the data on hand.

Aboard the National Oceanic and Atmospheric Administration ship *Discoverer* I saw both new and old methods used. I inspected new deep-current meters as well as devices that measure surface currents, water temperature, and salinity. Then I watched a technician take a wire and lower a dozen of the old Nansen bottles to collect water samples. When they reached their intended depths, a weight slid after them, triggering a mechanism that tipped them over to get their samples. Thermometers on the bottles recorded the temperatures.

As they came up again, the technician turned to me. "The new direct-reading gadgets are useful," he said, "but we still have to bring the water aboard for real precision." He took his samples to a laboratory to check their chemical properties, including oxygen content and salinity. The ratio of salts to water varies, and he had an instrument to measure it electrically.

With enough samples like these and a lot of math, scientists can plot the currents. Their flow is controlled by the earth's rotation, winds, and differences in density of water masses. Density is determined by temperature and salinity.

As water gets saltier, it grows denser, or heavier. Adding two tablespoons of salt to a quart of fresh water gives the average "saltiness" of the open oceans; but rain and rivers dilute many seas, and evaporation makes areas like the Red and Sargasso Seas more saline.

Salinity varies slightly in the oceans; even greater differences in density are caused by temperature because it varies far more — from about 28.6° F. near the poles, to 133° in a small area of the Red Sea. It averages 65° in the Northern Hemisphere and 61° in the Southern. Frigid levels of

arctic and antarctic water inch along the sea floor. Oceanographers who based their findings on these facts considered them woefully inadequate, for most charts represented averages only.

The Gulf Stream, giant river of the sea, was the first great circulation to yield secrets to a modern approach. In 1958 Henry Stommel, a specialist in ocean circulation, described to me a then radical theory based on the concept that differences in water density and not wind cause current movement in the depths. He predicted a strong southward-flowing countercurrent under the Gulf Stream.

To check Stommel's idea, Dr. John C. Swallow used a ten-foot aluminum probe he had developed. "We can set this float to remain within a hundred meters of any depth we choose," he explained. "It sends out an identifying ping." Dropping these pingers into deep water off South Carolina, he adjusted their buoyancy for various

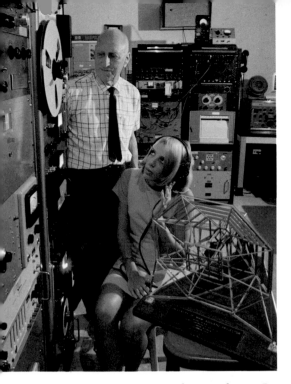

depths. Between 8,400 and 9,600 feet, seven pingers moved south, proving Stommel's countercurrent theory.

Since then, oceanographers have begun to probe ocean circulation in a variety of ways. Surface features are tagged for monitoring aircraft by floating buoys and marking dyes. And with infrared (heat) sensors, satellites can distinguish water masses directly, as in the case of large eddies breaking off from the Gulf Stream. Electronic sensors "read" temperature and salinity centimeter by centimeter and minute by minute to map the fluctuating patterns characterizing the waters of the seas. If a network of buoys carrying such sensors could be set up throughout the ocean, marine scientists could keep as close a watch on deep sea circulations as weathermen now keep on circulations of the atmosphere.

But such observations require much money. Even in the 1970's—the International Decade of Ocean Exploration—

Sounds of the sea intrigue a data analyst as Dr. John C. Steinberg plays a recording from an underwater acoustic projector (below). A model of the transmitter rests near them, in the Acoustic Laboratory of the Institute of Marine Science.

JERRY GREENBERG

current specialists complain about restricted funding, despite recent increases in appropriations for marine scientists.

Meanwhile, marine biologists see a different kind of threat in their work. With an eye on the sea's potential to help feed a hungry world, they hurry to learn more about the ocean's vast interacting patterns of life. They see these threatened by man's spreading pollution.

WATER is the preeminent substance of our kind of life," says Roger Revelle. "We now believe the ocean was the womb that first nourished and protected earth's children." And they occur nowhere in such giddy profusion as in the sea. Countless billions of microscopic plants stand at the very beginning of the ocean's long food chain, for they feed other countless billions of microscopic animals. Together, these tiny creatures make up a teeming community known as plankton.

Sir Alister Hardy, Professor Emeritus of Zoology at Oxford University, has written that plankton may represent "an assemblage of animals more diverse and more comprehensive than is to be seen in any other realm of life."

One day at Oxford, Sir Alister slid a plankton sample under a microscope for me. Diatoms, one of the most numerous of the plankton, glistened from their transparent silica shells like gems in little crystal caskets. Some, joined in long chains, shone like diamond bracelets. The diatoms share the ocean meadows with a second large group of plants, called flagellates. With their whiplike flagella, they keep themselves from sinking out of the sunlit upper water, where they live by photosynthesis.

Today the ocean womb, on which all marine life depends, seems dangerously vulnerable to man's pollution. Alarmed by what he sees in his explorations, Jacques-Yves Cousteau says emphatically, "The ocean is dying." He sees ecological danger in overfishing, which he calls "systematic destruction." Yet this could be halted by government agreement. The damage from pollution, harder to control, could quickly become irreparable. It threatens the plankton which sustain marine life and supply much of the world's oxygen.

Pollutants drain from the land, wash down with the rains, and ride windborne dusts. The top hundred meters of the sea contains ten times the lead—from car exhausts—it did half a century ago, not to mention ubiquitous oil and pesticides.

Professor Edward D. Goldberg of Scripps showed me data by which he can trace man's dirty finger prints all over the sea. In air samples collected over the Bay of Bengal, he found DDT residues probably brought by winds from as far away as Africa. He estimates windborne pesticide accumulates in the Bay at a rate of three tons a year. He notes that natural erosion sends about 3,000 tons of mercury a year to the sea. Man, he says, adds another 4,000 tons himself. Car exhaust from the United States alone contributes one tenth as much lead to the sea as do natural processes around the world. Man's influence on the sea thus approaches that of a force of nature.

While no scientist can clearly see what this influence portends, many share Captain Cousteau's concern for the ocean's health. "Worldwide draconian measures are called for if we want to survive," he says, "because it has been demonstrated that mankind cannot survive if the sea dies."

Meanwhile, the seagoing scientists carry on their work, and it yields many practical benefits. Biological explorations open new fisheries. Geophysical probing defines new oil fields. Chemicals from shellfish, sponges, and other creatures promise new wonder drugs. Indeed, the ultimate practical rewards of marine research defy prediction. But they may all slip through man's fingers unless he learns to approach the sea with a sense of responsibility as well as expectation.

Underwater TV camera sends sights and sounds of sea life to monitors a mile away on Bimini Island. The unit, controlled from shore, can rotate 360° and tilt up and down. A once-a-month scrubbing removes algae from its plexiglass dome.

4

Diving for Sport and for Science

BY BILL BARADA

Roaming a coral reef off Hawaii, amateur divers hunt marine life with cameras and spearguns. Such part-time adventurers have assisted scientists by inventing, testing, and developing underwater equipment.

SLIPPING INTO THE WATERS of the Pacific at Rangiroa Atoll, I flippered my way into the larger of two submarine passes that split the reef. Cautiously, I began to creep down one of its coral walls. Beside me, my friends Al Giddings and Dewey Bergman moved with the same discretion. The reasons for our vigilance were all around us—throngs of foraging sharks, including grays, black tips, and white tips.

We had planned this hazardous undersea itinerary with the hope of making a film documentary on sharks in a "feeding frenzy." Like many divers before us, we sought to pierce some of the riddles of shark behavior by observing and recording it at close quarters. We had found a perfect studio here in the Tuamotu Archipelago. In this cleft in the coral, hundreds of five- to ten-foot sharks prowl a 300-yard-wide, mile-long strip of water.

As the darkly ominous shapes swarmed about us, Dewey and Al got set with their movie cameras. I had a still camera, but my main job was to shoot fish with my speargun, attracting sharks to the bait and sending them into action.

The sharks began swimming quietly on all sides of us, keeping a distance of about 15 feet—too far for good filming. Each time I aimed the speargun, one shark would line up behind my target. I could see his eyes rolling as he waited for me to fire. Excited, the others moved swiftly around him. When a 15-pound snapper swam into range between the cameras, I shot it. The sharks went wild, coming at us from all directions as they raced for the fish.

Dodging the voracious predators, Al and Dewey swept their cameras over the mad scene while I gripped the harpoon line with one hand and held myself fast to the coral with the other. When a shark hit the fish on the end of the line and took off, a pack of about 50 zoomed into a tight circle over our heads. The piece of coral I clung to snapped off, and the thrashing brutes fighting over the fish on the end of the line began pulling me toward them. I let go of the line and watched gratefully as the battling pack swam away with the prize.

THE AUTHOR: *Bill Barada, former marketing manager for* Skin Diver *magazine and developer of underwater equipment, began diving in 1935 when the sport was called "goggle fishing." In 1940 he formed the Los Angeles Neptunes, one of the first diving clubs in the United States. Now a full time free-lance writer, he is the author of six books on diving, and has contributed numerous articles to national magazines.*

Five days and 2,500 feet of film later, we concluded our dives among these dreaded creatures. We had found that sharks don't always live up to their reputation of being dangerous man-eaters—at least not until familiarity really starts breeding contempt. Each day they became more aggressive, and we had a number of very close brushes.

Why did they tolerate us at all? Was it the fresh seafood dinners we provided? Perhaps. As investigators know so well, sharks react unpredictably to the presence of human beings, but most often attack surface swimmers or those who have been injured or appear distressed. Our self-contained underwater breathing apparatus, or scuba, let us move about among them with ease and control.

Marine biologists who have seen the film tell us it represents an important addition to the archives on shark behavior. If so, it will be only one of countless contributions made by amateur divers to the fund of knowledge about the sea and its denizens.

David R. Stith, former president of the Underwater Society of America, agrees with me that sports divers are playing a far bigger role in underwater developments than anyone anticipated. As Dave put it: "They have come up with a lot of invaluable information, techniques, and improved equipment. And if the confident amateur hadn't led the way, many of the marine scientists and engineers who are now getting wet might have taken much longer to use scuba as the vital research tool it is."

The art of diving had progressed unbelievably between my shark dive in 1966 and the day in 1935 when, with only a homemade spear and goggles, I first peered into the undersea world off Palos Verdes Peninsula, California. I had been spurred to this adventure by hearing that the world's

Its four-foot-long arms flailing, a Pacific octopus (Octopus dofleini) *struggles to escape the firm grip of William L. High as he drags it from its anemone-covered lair in Puget Sound. At the approach of man, the naturally timid creature usually tries to hide or to camouflage itself by changing the color and texture of its skin to match its surroundings. Wrestling the 35-pound*

NATIONAL GEOGRAPHIC PHOTOGRAPHER BATES LITTLEHALES

cephalopod to shore, High, a biologist for the National Marine Fisheries Service, will use a specially designed hypodermic needle to insert plastic tags into the base of one of its arms and its body. Adults of this species weigh 80 pounds or more. Tags identify individual invertebrates so High can study migration patterns in an effort to learn more about the life cycle of the octopus, one of the most intelligent cold-blooded ocean dwellers. In addition to this project, begun in 1957, High represents the United States on a United Nations Food and Agriculture Organization committee for submersibles and undersea habitats. He also teaches scuba diving at Shoreline Community College near his home in Seattle, Washington.

73

Census takers with plastic slates and grease pencils count Florida Keys reef fishes for the American Littoral Society, an aquatic study and conservation group. A Sandy Hook Marine Laboratory diver collects swaying current detectors off New Jersey. Above, an amateur diver with a slurp gun captures wrasses for his home aquarium by drawing them into a water-filled chamber.

first skin-diving club, the Bottom Scratchers of San Diego, had formed in 1933. And Guy Gilpatric had inflamed my imagination with descriptions of spearfishing in the Mediterranean in a magazine article, "The Compleat Goggler," which later became the title of his famous book. But nothing had prepared me for the reality of my first goggle dive. Mesmerized, I floated above a breathtaking seascape of color and movement. A jungle of kelp and seaweed undulated with the swells. Jagged rocks bristled with dazzling arrays of clinging organisms. Weaving through kelp and peeking out from holes in the rocks, garibaldi goldfish,

perch, and rockfish kept a wary eye on me.

Several times I dived to about 20 feet and learned the limitations of my crude equipment. The goggles pressed tightly against my eyes, the spear impeded my progress, and almost as soon as I touched bottom my lungs ached for want of air and I hastened back to the surface. Within 15 minutes I "froze out" from the chill of the water and returned to shore. But I firmly resolved to continue exploring this enchanting world.

My diving friends and I faced the constant frustration of inadequate equipment. At first, we improvised goggles. Later we chanced upon some face masks imported

from Japan. These suggested ways of making our own, and we greatly improved our range of vision as well as our enjoyment of the beauty below.

In 1940, Olympic yachtsman Owen P. Churchill of Los Angeles introduced rubber swim fins into the United States. He told me that he had seen divers in Tahiti wearing them. When he found they had been developed by French Navy Comdr. Louis de Corlieu, he bought the patent rights for the United States market. These flippers increased our speed and range. By taking the work out of swimming, the fins allowed amateurs to dive like experts.

Then veterans of World War II underwater demolition teams brought home their rubber frogman suits. A member of my

Speargun poised, ichthyologist Eugenie Clark searches for reef fish among corals and sea fans near Grand Cayman Island in the Caribbean. Like many marine biologists, she began diving to survey the underwater world firsthand.

first developed and manufactured by an amateur. Air trapped inside the cellular foam rubber or neoprene of a wet suit provides insulation. Water seeping under the suit is kept warm by body heat.

When diving suits broke the cold-water barrier in the late 1940's, new clubs sprang up under colorful names — the Boston Sea Rovers, Coast of Maine Neptunes, and Puget Sound Mudsharks. This mask-and-fin, hold-your-breath period saw divers taking great pride in exhibiting their prowess, stamina, and courage. Spearfishing gave ample opportunity for such expression, and competitive meets became the vogue.

For some time we had been using the snorkel, a tube permitting us to breathe while face down on the surface. But every diver dreamed of obtaining a simple, safe device for breathing underwater. The oxygen breathing gear used by Navy frogmen did not qualify for amateurs. It could bring disaster to the uninformed, because pure oxygen becomes a deadly poison if divers go too deep.

The dream began to materialize in 1949 when a California sporting-goods dealer imported the Aqua-Lung. Interest should have exploded like a bomb, but it didn't. For one thing, a basic set of scuba, including air tank, cost $250, even more than today's price. Most dealers knew nothing about diving and less about the Aqua-Lung, since the first directions for its use were in French. And for a time no air stations existed to recharge the tanks.

I acquired my first Aqua-Lung in 1950 by trading the importer three rubber suits I had developed. He gave me oversimplified instructions, to say the least: "You know how to dive. Just put on your 'lung' and go ahead. It's automatic." I loaded my new equipment into a rubber boat and with my wife, Harriet, headed for the

club, the Los Angeles Neptunes, obtained one, and fellow divers immediately began making copies, cementing them together in garage workshops. Protecting us from the cold and keeping us dry, the suits increased our endurance time, and we could enjoy hours of diving in comfort.

Now divers can choose between ready-made "dry" suits and "wet" suits, the latter

Captive killer whale, Namu permits inspection of his formidable array of teeth by his owner, Edward I. Griffin of Seattle. The 24-foot-long relative of the porpoise, seeking food and *companionship, glides under Griffin's skiff and overturns it, spilling his keeper and a load of salmon. Namu's playfulness contradicts the belief that these whales are invariably killers.*

Five-ton Namu cavorts in Rich Cove near Seattle after eating part of his daily ration of 400 pounds of salmon. Griffin towed his Orcinus orca in an underwater pen from British Colum- *bia, where a fishing net had accidentally snared the whale. Prior to Namu's death in 1966, scientists listened to his heartbeat and recorded his sonarlike beeps and squeals.*

Straddling Namu's broad back for a 15-minute romp, Griffin clutches the high dorsal fin for balance. "So sensitive is the fin," he said, "that at first the touch of a finger alarmed Namu, and he would quickly shake me off." The bull whale soon became so accustomed to his companion that at times he slept with Griffin aboard.

kelp beds off Point Dume, near Malibu.

The use of quick-release buckles and slip hitches that allow instant removal of a tank —a basic safety rule of present-day scuba instruction—was unknown then. I tightly strapped on my tank so it couldn't possibly work loose, then slipped into the water.

It took only a few minutes to adjust to the Aqua-Lung before I felt an exhilarating sense of freedom. I breathed as naturally underwater as on the surface! Moving like a fish, I swam through stands of giant kelp, hovered above a rocky pinnacle, and lingered along the face of a cliff. I drifted down, gliding into depths I had never reached before and quickly found an abundance of abalone, lobsters, and fish to fill the sack I carried. All too soon my air gave out and I started to ascend.

On the surface I ran into trouble. I came up in a heavy kelp bed, 50 yards from the boat. Water that had leaked into my suit and 16 pounds of lead in my belt weighted me down. Although accustomed to holding my breath and swimming beneath the kelp canopy, I had never reckoned with an air tank in this circumstance. My tank, fouled in the kelp and partly out of water, became heavier. Gasping for air and struggling to stay afloat, I fumbled with the straps of the Aqua-Lung and the weight belt. If Harriet had not reached me with the boat, it could have been my last dive.

When Jacques-Yves Cousteau's remarkable book of diving adventures, *The Silent World,* appeared in 1953, Aqua-Lung sales began to go up in the United States. The boom continued as the public watched underwater motion pictures and television shows. But knowledge about Aqua-Lungs failed to increase at an equal pace, and untrained divers sometimes met with disaster.

Dolphin snatches a fish nearly 20 feet above a portion of Florida Bay fenced in for Flipper's Sea School at Grassy Key, near Marathon. Using a type of sonar, trained dolphins have great ability to locate and retrieve objects underwater. Divers on occasion use them to carry messages and to locate lost companions.

A typical example of the attitude that prevailed occurred during a club dive off Santa Catalina Island. Three strangers, each with a full set of new equipment, joined us aboard a chartered boat. I discovered none had ever dived before. When I suggested they practice with mask, fins, and snorkel before putting on their tanks, one said, "We can't do that. We have to be able to breathe because none of us can swim." After I told them the facts of scuba diving they decided to stay on the boat.

ANTICIPATING THE THREAT of restrictive legislation and anxious to avoid it by establishing their own controls, diving clubs began forming regional councils. In 1959 the councils joined together as the Underwater Society of America. The society in turn became the largest constituent of the World Underwater Federation, established in Paris later that same year.

When *Skin Diver* magazine first appeared in 1951, its pages informed divers of problems besetting their sport and offered suggestions for finding solutions. The search for means to ensure safety began in earnest on the West Coast.

Bevly B. Morgan, then a lifeguard with the Los Angeles County Department of Parks and Recreation, became concerned over the growing number of rescues involving scuba divers. He advised county officials to enlist the services of experts in setting up classes on diving safety.

The expert divers who were invited included two graduate students at the Scripps Institution of Oceanography—the late Conrad Limbaugh and Andreas B. Rechnitzer. Both went on to win recognition as marine scientists. Their group produced the first book of safety rules for divers and launched a program to certify instructors.

By 1971, the Los Angeles County program had certified more than 600 instructors who had trained some 130,000 skin divers and scuba divers. Clubs around the country have developed programs patterned after the Los Angeles operation.

Meanwhile, the Young Men's Christian Association, in whose pools many swimmers

Natural friends of man, dolphins tow a companion at Sea Life Park near Honolulu, Hawaii. Above, frisking mammals obey a command, beeped to them by a battery-powered signaler, to stay near the boat. Below, scientists record the heartbeat and muscular activity of an anesthetized research animal in a Miami laboratory.

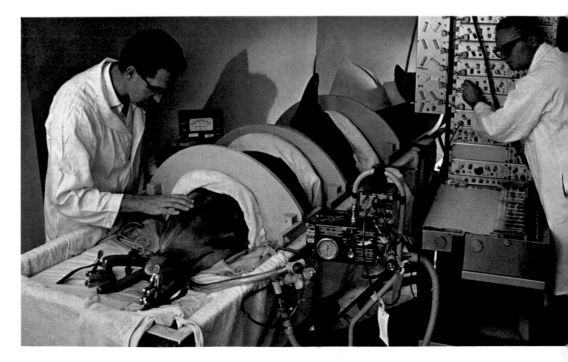

were learning scuba, worked with the Conference for National Cooperation in Aquatics to prepare a training manual in 1957. Classes throughout the United States use it. In 1959 the YMCA began a nationwide project to certify instructors to help ensure uniform requirements.

Around the globe, diving groups took similar steps in the interest of safety. Diving magazines further educated their readers. Through its official magazine, *Triton*, the British Sub-Aqua Club publicized its diving course, regarded as one of the world's best. So effectively did the Sub-Aqua Club police the sport that Parliament decided not to require licenses for divers.

Just how safe is the sport now?

John J. McAniff, director of the National Underwater Accident Data Center at the University of Rhode Island, points out that the number of active scuba divers in the United States increased rapidly in 1972, raising conservative estimates of the total to more than 500,000. "Yet the 122 diving fatalities reported that year represent an increase of only 2.6 percent over the year before," he said. "So it would appear that diving is becoming safer. Analyzing the causes of both fatal and nonfatal underwater accidents helps to increase safety by contributing to improvements in instruction, practices, and equipment."

In the early days, spearfishing attracted most scuba divers. But as the novelty wore off many of them started looking for underwater activities with a purpose. As a consequence, marine archeology, salvage operations, photography, engineering, and biology owe a great deal to amateur divers.

For example, along the Pacific coast from San Francisco to Mexico, divers have found thousands of artifacts left by prehistoric

seashore inhabitants. Just off the La Jolla Beach and Tennis Club, divers have picked up hundreds of mortars and other stone objects, proving the existence of villages beyond the present shoreline as long ago as 7,000 years.

On the Atlantic coast the divers' search for useful activities has had significant results. One day in 1960 in the office of Dr. Lionel A. Walford, former director of Sandy Hook Marine Laboratory in New Jersey, "a group of divers appeared and demanded something to do."

He put them to work on a number of projects. "From this nucleus of enthusiastic volunteers came the American Littoral Society, organized in 1961," Dr. Walford said. "As amateur naturalists they have accumulated much new information about fish. Of course, they've needed guidance and training, and we're happy to give it to them when time permits."

A dozen years later the society had 4,000 members, some in nearly every state. About half are divers. They carry out various assignments, reporting their findings to the society: counting and tagging fish; collecting data on plant and animal life in water affected by nuclear power plants or sewage; observing marine life in artificial reefs. Nixon Griffis, society president, mentioned Arizona divers' help in measuring ecological changes in the Colorado River estuary.

"Irrigation projects and dams increase river salinity and reduce flow. This upsets the delicate ecological balance, impairing the estuary as a spawning area for certain fish. For example, divers' reports indicate that the number of totoava, a valuable Gulf of California sport and food fish, has dropped dramatically in recent years."

According to society member David K. Bulloch of Hillsdale, New Jersey, "Watching undersea creatures gets under your skin. Curiosity brings you back to the same spot week after week to see what has happened to the inhabitants."

He should know. A research chemist, Dave devoted his spare time to observing invertebrates living on a wreck 82 feet

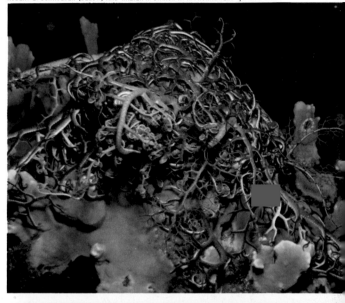

Wearing self-contained underwater breathing apparatus, or scuba, a diver hitches a ride on a loggerhead turtle off the Florida Keys. The huge reptile, its jaws powerful enough to mangle an arm, tries to lose its passenger by plunging to the bottom. In darkness (lower right), the Gorgon's head, or basket star, an animal that resembles a plant, stretches ghostly arms to trap plankton. Crumpling into a matted heap (upper right), the eerie creature assumes its daytime posture.

NATIONAL GEOGRAPHIC PHOTOGRAPHER OTIS IMBODEN (BELOW)
AND WALTER A. STARCK II

Emerging from Florida waters, marine biologist Walter Starck hands his wife, Jo, a camera that he adapted for his ecological studies. The plastic dome, covering a Fisheye lens, helps sharpen its focus during underwater use. At right, the lens frames Mrs. Starck among large clumps of brain and star corals off Lower Matecumbe Key.

down in the Atlantic off Shark River Inlet, New Jersey. His report in *Underwater Naturalist,* the society's quarterly, brought high praise from biologists.

In the early 1950's, Conrad Limbaugh sounded the first call for scientists to use scuba in studying the marine environment. Soon afterward, David M. Owen promoted the same idea at Woods Hole Oceanographic Institution. Today hundreds of scientists go down into the ocean to observe and collect, employing equipment and techniques developed by amateur divers.

In 1959 three of California's original group of diving scientists, Limbaugh, Dr. Wheeler J. North, and James R. Stewart, swimming into a sea canyon during a storm, became the first to report one of nature's unsuspected underwater tricks. At Cabo

San Lucas, Baja California, they found sand falls, which resemble waterfalls with the materials switched around—land plummets through water rather than water through land.

At 30 feet, the divers noticed sand sliding gently down the canyon slope. Following the flow, they swam above a dozen or more falls tumbling among the rocks. Finally, when they reached 130 feet, they saw a spectacular stream of white sand cascading over a sharp dropoff. The storm waves on the surface were cutting sand from the beach and driving it into the canyon. In later descents, Dr. Robert F. Dill swam along sand falls and rapids at 250 feet. From that level he could see the sand river flowing a hundred feet farther down.

Dr. Eugenie Clark of the University of Maryland can be found underwater as often as on land, collecting specimens in jars and nets. While diving near Eilat, Israel, she discovered a new species of *Trichonotus,* a small, elongated fish commonly known as a sand diver. Following scientific custom and a mother's prerogative, she named the fish *Trichonotus nikii,* in honor of her son Niki.

Between 1963 and 1966, marine biologist Carleton Ray and his associates braved frigid waters to observe great Weddell seals under the Antarctic ice. "We would be going down with standard breathing equipment," Dr. Ray reported in the NATIONAL GEOGRAPHIC, "but our suits had been specially fabricated for us of 5/16-inch-thick foam neoprene." The only exposed skin was a little area around the mouth, which the men knew from experience would not get uncomfortably cold in icy water.

On one of their first Antarctic dives in 1963, Dr. Ray and Navy Lt. David Lavallee encountered a nine-foot, 800-pound seal heading for the same hole in the ice that was their objective. "Almost together the seal and I popped our heads through the access hole," Dr. Ray said. "On an impulse, I gave him a gentle pat on the snout, and then heaved myself out of the water.

"Dave and I watched the animal breathe deeply and repeatedly for three minutes, his nostrils dilating and contracting like

giant mechanical valves. Before he clamped his nostrils shut and dived again, he gave us one brief glance and that was all.

"I couldn't have been more delighted by his casual acceptance of us. . . . To him we were fellow marine creatures, and he was proving our motto: 'If you want to study a seal, *be* a seal.' "

Under these intimate circumstances, Dr. Ray and his colleagues watched seals be-

Amid gnarled antler coral jutting from the sea floor, a swimmer ranges the underwater trail in Buck Island Reef National Monument, U. S. Virgin Islands. In Florida's John Pennekamp Coral Reef State Park, the nine-foot statue "Christ of the Deep" stands in 24 feet of water.

neath the ice snoozing while holding their breath and also listened to seal "talk." "There was not a moment's silence in the 'silent world,' " Dr. Ray wrote, "but instead a never-ending chorus of seal trills, chirps, and whistles."

Dr. Walter A. Starck II, another diving marine biologist, started using an Aqua-Lung at 14 and soon began taking pictures with a camera in an improvised rubber-bag housing. These beginnings eventually led him to embark on one of the most far-reaching studies of coral-reef life ever made.

About 1958 Walt selected Alligator Reef off the Florida Keys for intensive research. "The Florida coral reefs were about as well studied as any in the world at the time my

wife, Jo, and I started diving there," he told me. "But nobody had ever really gone down into the water to see what went on. Scuba let us do just that."

With support from the National Geographic Society and the National Science Foundation, Walt and Jo found 517 species of fish on Alligator Reef—the richest known fish fauna of any area in the New World. "Forty-five of those species were previously unknown in United States waters," he said, "and 18 were completely unknown to science. Gives you an idea of how much we really knew about the reefs before, doesn't it?"

Increasing threats to marine life in coral reefs from pollution, careless spearfishing, and thoughtless coral collectors have prompted conservationists around the world to begin staking off preserves similar to those protecting game and plants in the great national parks. The underwater preserves offer opportunities for both recreation and scientific study.

The first, and one of the most beautiful —John Pennekamp Coral Reef State Park in the Key Largo reefs of Florida—came into being in 1960. Here snorkelers and scuba divers may explore a 21-mile-long submarine coral garden. Protected from shell collectors and spearfishermen, sea life abounds. The fish, like animals in the parks, show no fear of man. Swimming around a diver's head, they peer boldly into his face mask, begging for handouts.

The Japanese Nature Conservation Society is investigating a number of sites for possible use as marine preserves. In France, the famous diver and underwater archeologist, Philippe Tailliez, has played a leading role in establishing the Port Cros Submarine Park near Toulon.

Perhaps the most popular spot for underwater sightseeing, the U. S. Virgin Islands now teem with visitors rambling over and among the coral reefs. The Department of the Interior maintains trails in marine parks off Buck Island and St. John.

Snorkelers, swimming at the surface, look down at submerged markers 10 to 20 feet below that identify the various types of coral and explain their formation. Some signs show pictures of fish, give their names, and inform the waterborne sightseer about their habits. Even nonswimmers with snorkels can glide above the coral, holding on to floats towed by small craft.

Today the camera replaces the speargun as members of the "wet jet set" use the oceans as their playground. In the two decades following my first diving experience, diving became as commonplace as skiing.

Occasionally I feel a touch of nostalgia for those early days when a handful of adventurers probed the mysterious submerged world. More often, though, I remember with pride that the adventurous sports diver prepared the way for an even more exciting era of scientific discovery.

Sails furled, boats hover above their shadows in Buck Island's turquoise-green lagoon. Sloops and catamarans ferry snorkel swimmers and scuba divers to the barrier reef from nearby St. Croix. In Pennekamp Park, an underwater visitor offers a bit of sea urchin to a venturous white grunt. Tiny bluehead wrasses wait for leftovers.

5

Cameras Below

BY LUIS MARDEN

Like performers awaiting a cue, striped grunts cluster in shallow waters off Bimini, in the Bahamas. Despite poor visibility and limited mobility, today's underwater photographers match their surface colleagues in capturing photographs of great clarity and beauty.

SHORTLY BEFORE the United States entered World War II, I went to the Antilles to write an article for the NATIONAL GEOGRAPHIC on U. S. military bases in the islands. Between bomber flights I stayed for some days on Antigua, and there I had my first look at a coral reef through a diving mask borrowed from a Marine Corps officer.

Like almost everyone else who has known the experience, I was entranced. The eerie beauty and strangeness, the other-worldly landscape and alien life forms, seemed to belong to another planet. Making pictures is part of my profession, and I yearned to photograph the scene. Unhappily, on an island in wartime, I had no way of improvising an underwater camera. So I simply observed and marveled as I explored this mysterious liquid world.

I did not know it at the time, but half a century before, a Frenchman, Louis Boutan, had anticipated my experience.

". . . with regard to the sea," he wrote in the first book on underwater photography, "the naturalist is like an inhabitant of the moon sailing through ethereal space, who cannot descend through the atmosphere that surrounds the earth.

"If he wishes to obtain some notion . . . of the globe and its inhabitants, he would have to . . . construct dredges . . . to make contact with the surface of the earth.

"His nets might capture some birds, which would represent for him the most numerous inhabitants of the earth, and, if his dredge were to knock off the top of some factory chimney, he would conclude that it was the mysterious abode of some unknown animal.

"Up to the present, naturalists have operated exactly in this manner in their study of the ocean deeps . . . blindly.

"How the situation would change the moment it becomes possible to take photographs on the bottom of the sea!"

Boutan wrote this in 1900, after he had worked on the problem of the submersible camera for eight years. Almost singlehandedly, he created the art of making pictures underwater. A marine biologist of robust build and questing mind, he made his first descent in helmet and diving dress in July 1892 in the Mediterranean.

He recalled: "The strangeness of the submarine landscape made a strong impression on me.... I then resolved to try photography. If one can photograph a landscape in open air, why, I asked myself, should it not be possible to make a photograph on the bottom of the sea?... there should not be any invincible obstacle...."

He constantly improved his equipment and with his third camera, a great copper and iron box weighing hundreds of pounds, he made his best pictures.

The camera had to be lowered from a boat to Boutan, who wrestled it into position. He wrote: "Even though the camera,

THE AUTHOR: *Luis Marden, Chief of* NATIONAL GEOGRAPHIC'*s Foreign Staff, has photographed the ocean world for 30 years. An early specialist in 35mm color work, he published the first book on the subject,* Color Photography with a Miniature Camera, *in 1934. In 1957 he discovered and photographed the remains of Captain Bligh's* Bounty *off Pitcairn Island.*

lighter when submerged, could be handled by one man under water... many times I have worn myself out... the sweat rolled down my forehead; the steam... condensed on the windows of the diving dress, and the landscape was barely perceptible through a dense fog.

"Not having the use of my hands within the helmet, I could do nothing but rub the ... glass with my nose or my tongue... to make a small window through which I could perceive objects more clearly."

Poor perspiring Boutan later made things a bit easier for himself by using an empty wine barrel as a float and suspending the camera from it.

With the slow photographic plates of his day, Boutan found it difficult to make "instantaneous" photographs—snapshots. At best he could only squeeze the shutter open and shut for about 1/50 of a second, barely enough to stop the motion of gorgonians swaying in the current. In despair, he tried a new lamp designed for him by an electrical engineer, one M. Chaufour.

LOUIS BOUTAN, "LA PHOTOGRAPHIE SOUS-MARINE," 1900 (BELOW) AND MARITIME MUSEUM OF PHILADELPHIA (OPPOSITE)

Father of undersea photography, Louis Boutan takes a self-portrait 12 feet down. An empty wine cask buoys his heavy camera (left). In 1892 he photographed a Mediterranean spider crab (opposite), in the first known underwater picture.

94

Patterned hogfish roams sandy shallows of the Gulf of Mexico in a photograph from the world's first set of underwater color pictures, made in 1926 by ichthyologist W. H. Longley and Charles Martin of the National Geographic Society.

Throughout its history, the Society has supported experiments in undersea photography. In a 1968 photograph by Dr. Walter A. Starck II, a seldom seen blackcap basslet (Gramma melacara) noses between coral branches of a reef off Andros Island.

It burned a twisted ribbon of magnesium, ignited by a platinum filament that glowed when battery contacts were connected. The magnesium burned unevenly and Boutan gave up the lamp, but the ingenious device was the ancestor of the flashbulb.

Finally, in 1899, after unsuccessful experiments with another lamp, the resourceful Parisian developed an unmanned deep-sea camera, the chief photographic tool of oceanographers today. In glass-and-metal spheres on each side of the housing, he placed arc lights powered by batteries on the vessel above. Opening the shutter by remote control, he made a sharp picture at 160 feet.

Boutan concluded his experiments in 1900. At the end of his book he wrote: ". . . I have opened the way . . . it remains for others to follow me, to open up new paths, and to arrive at ultimate success."

In 1914 English-born J. E. Williamson made the first undersea motion pictures. From a barge he hung a flexible metal tube; through it a man could descend to a steel sphere 30 feet below. Inside the capsule, big enough to hold two men with cameras, dry divers photographed the sea life of Bahama coral reefs through a plate-glass window. From this "photosphere" Williamson made black-and-white and color films.

But the first man to make an extensive series of still pictures of fish in their natural habitat was, so far as I know, Professor William H. Longley, an ichthyologist at Goucher College in Baltimore.

Professor Longley walked the sea bed off the Dry Tortugas, a coral islet beyond Key West, Florida, in 1917. He used a Graflex camera in a beautifully made watertight brass box that I saw many years later in a laboratory of the Smithsonian Institution.

The professor made hundreds of black-and-white photographs, showing for the first time the habits of many fish: a trunkfish blowing sand away from its food, a trumpetfish hiding head down in the plumes of a sea feather, and a host of other things that cannot be learned from shriveled brown specimens preserved in alcohol.

When Dr. Longley submitted an article on life of the coral reef to the NATIONAL GEOGRAPHIC in 1926, the Magazine's editors, who had pioneered in the use of color photographs, thought at once of undersea pictures in color. But the color plates then used by the Society, the Lumière Autochrome, were exceedingly slow. They required a one-second exposure at f/8 even in bright sunlight, and it seemed impossible to photograph moving objects in the diffused undersea light.

AT THAT TIME, Charles Martin, an ingenious technician and innovator, headed the National Geographic's Photographic Laboratory. In July of 1926, the Society sent him to Dry Tortugas to collaborate with Dr. Longley. The two soon discovered that, like Boutan, they needed artificial light, lots of it.

More light and moving subjects meant flashlight photography. Before the invention of flashbulbs, flash pictures were made by igniting powdered magnesium with a spark. The quantity normally used was about one ounce, but to record submarine scenes on the Autochrome plates, Martin used an incredible *one pound* per flash, exploded on a raft.

In a letter to the Society, Longley wrote: "I hope we may succeed gloriously, but it is a gamble." It was also dangerous. The powder went off with a blinding flash in a cloud of smoke, lighting the sea bottom with the equivalent of 2,400 flashbulbs, more light than has ever been used, before or since, to make pictures under the sea.

There was no precedent for this kind of photography; Charles Martin had to invent or improvise all his equipment. He built a pontoon raft to carry the powder and the apparatus for firing it. Over the raft he stretched a white cloth to reflect the light downward. Dr. Longley, walking 10 to 15 feet below, towed the light raft about as he moved. Martin even devised the synchronizer that would fire the powder when the camera shutter opened.

When he returned to Washington, he could report triumphantly: "Herewith are

JERRY GREENBERG (BELOW) AND EDMOND L. FISHER (PLANKTON)

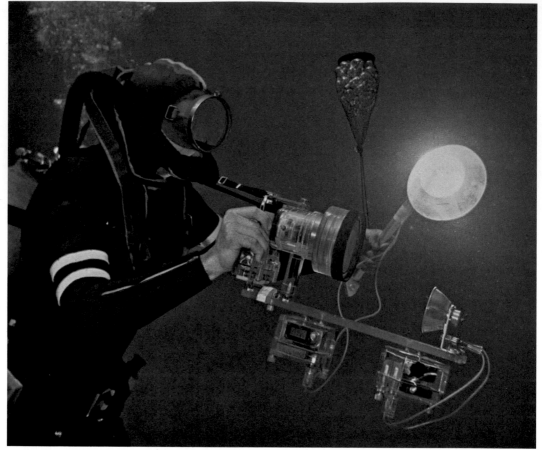

Firing a Nikonos camera with hand-held and frame-mounted flashes, a photographer takes portraits of plankton, the oceans' tiniest inhabitants, with a close-up lens. Spare flashbulbs float in the net bag. Drifters more than swimmers, plants as well as animals, plankton comprise the basic foodstuff of the sea. Below, left to right: a section of a siphonophore, a jellyfish colony buoyed by gas-filled floats; trochophore, a peanut-worm larva; protozoan, a one-celled animal with tiny whips for beating slowly through the water; acorn-worm larva, a trochophore that spins around as it swims; sea butterfly, a "flying" mollusk, its streamer sensory tentacles extending from winglike feet; and a mollusk veliger larva wrapped in a crystalline membrane.

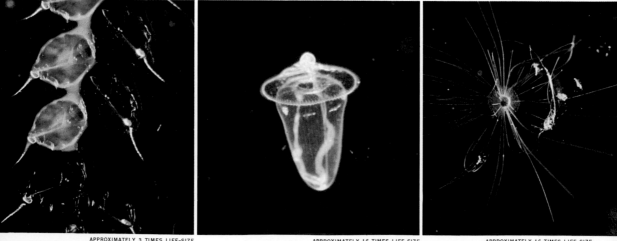

APPROXIMATELY 3 TIMES LIFE-SIZE   APPROXIMATELY 16 TIMES LIFE-SIZE   APPROXIMATELY 16 TIMES LIFE-SIZE

eight autochromes of genuine sub-marine life; the first ones ever taken."

They were also the first undersea color photographs ever published. They appeared in the January 1927 NATIONAL GEOGRAPHIC, and marked a milestone in the history of photography. Nearly 30 years passed before the publication of another undersea color photograph.

After World War II ended, the Aqua-Lung came on the world market, and increasing numbers of amateur divers began to go down into the sea. At once they tried to take cameras with them, to show the stay-ashores what the clamor was all about. The first problem was simply that of keeping the camera dry. All kinds of clever makeshifts appeared, from cameras encased in Mason jars to cameras looking through a glass port in rubber hotwater bottles. Only a few divers had the skill or means to attempt a proper case of metal or plastic through which camera controls of film advance, focus, iris diaphragm, and shutter speeds would be workable.

By that time I had obtained my first commercial underwater housing, a sea-green metal cylinder made in Venice. For some time I had been trying to make color photographs under the sea, but there was a hitch. Martin and Longley had taken their color pictures in crystalline water 10 to 15 feet deep. Lighted by that stupendous flash overhead, the brilliant color of the coral reef came through faithfully. But as depth increases, the thickening blue-green filter of seawater, interposed between the sunlight and the bottom, absorbs the colors of the spectrum. First the reds, then the oranges and yellows, disappear, until beyond 30 feet or so, the diver walks or swims in a monochromatic blue-green world.

In shallow clear water, the eye adjusts to the prevailing blue and can still see some reds and yellows, darkened and degraded from their surface brilliance. My first pictures disappointed me, because they looked like black-and-white photographs tinted blue-green. Unlike the human eye, the photographic emulsion does not possess the power of adaptation, and all the warm colors were drowned in blue light.

I used correcting filters to hold back the excessive blue, and got fairly good photographs in the limpid waters of the Mediterranean, provided I did not go too deep. But I soon learned that to make really good color pictures underwater I had to use an artificial light source close to the subject. Flashbulbs seemed ideal, and in 1955 when I accompanied Captain Cousteau aboard *Calypso* during his filming of *The Silent World*, I took along 600 of them.

Cousteau later recalled our experience: "Our divers wonder how a man can fire that many bulbs in only four months. They soon discover how. Hardly have we left the Suez Canal before Marden begins

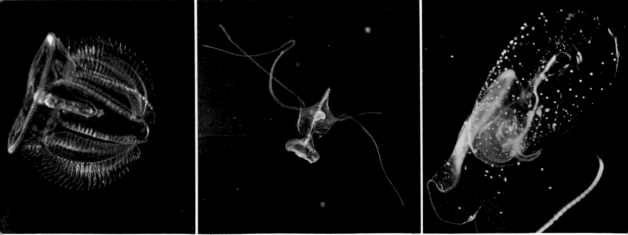

APPROXIMATELY 16 TIMES LIFE-SIZE   APPROXIMATELY 4 TIMES LIFE-SIZE   APPROXIMATELY 20 TIMES LIFE-SIZE

His camera protected by a special housing he designed, Dr. Starck takes a close-up photograph of

diving with stout Émile Robert as bearer.

"Robert goes down carrying Marden's second camera and a large string bag of flashbulbs, which floats above him like the envelope of an 18th-century balloon. It is not long before Marden is crying for more bulbs and we radio for a fresh supply.

"The bulbs, when under pressure, develop leaks in the metal bases. Water seeps inside, short-circuits the lead-in wires, and makes firing uncertain.

"Marden is chagrined, but the *Calypso* team comes to his rescue. At night we see a strange scene in the mess. The ship's cook

marine life on a western Caribbean reef.

wax into the holes to insulate the wires.

"Luis's expenditure of bulbs taxes the production rate: It is a race between manufacturer and consumer. We treat 2,500 bulbs before the voyage is over."

Most of them worked. But at depths of a hundred feet or so, something else happened: The bulbs sometimes imploded, shattering violently inward, instead of outward as in an explosion, and driving the fragments of glass into my gloved hand like bullets. The pure oxygen that fills flashbulbs is at a pressure less than one atmosphere, so their thin glass walls undergo strain even at relatively shallow depths. After firing weakened the glass by developing minute cracks in it, some bulbs imploded when touched.

This happened to me the first time as I swam above a sunken ship at 90 feet in the Red Sea. When I took hold of the bulb to remove it from the reflector, I heard a dull report and felt a numbing pain lance through my right thumb. A wisp of greenish "smoke" slowly curled upward; it was my blood, drained of its bright red at this depth.

Weeks later, when another implosion left a neat scar on the middle finger of my right hand, tracing the exact curve of the big bulb, I started to search for some kind of protective glove. But the stoutest leather was ineffective against the tremendous impact of the shattering glass. A chain-mail glove made for butchers provided the answer. Now I handle even the jagged bases of shattered bulbs with impunity.

It is an uncannily beautiful sight to look at a calm sea on a dark night when underwater photographers are at work. Suddenly and silently a circle of sea a hundred feet in diameter briefly flashes firefly green, like heat lightning on the horizon. Three-quarters of a century ago, Louis Boutan's skipper noted the awesome effect of the professor's submarine flashes. "It is as though a storm is raging under the sea," he said.

As a result of Cousteau's expedition I discovered another problem with color. Swimming 60 feet down in the Indian Ocean, I had seen a clump of sea anemones

heats water; the second cook melts wax in the water; my wife, Simone, cleans the bulb bases; the engineer drills two tiny holes in the base of each bulb; and at the end of the production line stands the ship's young surgeon in his white tunic. With the delicacy of a brain surgeon he injects liquid

RON TAYLOR

Savage jaws agape, a great white shark bares two-inch dagger-sharp teeth seconds before assaulting and devouring a dead shark tied to an Australian fishing boat. Diver Ron Taylor recorded the swift attack with a movie camera.

Drifting among oceanic whitetip sharks off South Africa (above), photographer Peter Gimbel ventures from a protective cage to film a nine-foot bull for his documentary movie Blue Water, White Death. *A great white shark (right), the object of the film crew's search, charges the shark cage off Australia—the only area where Gimbel found the great white although it has a world-wide range. The most deadly underwater predator and one of a dozen shark species listed as man-eaters, the great white shark attains an average length of 18 feet and a weight of more than 4,000 pounds; its jaws can spread three feet. An omnivorous eater, it prefers a diet of large vertebrates.*

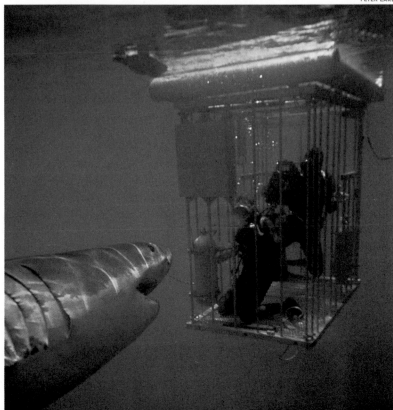

with globular tentacles like bunches of brilliant red grapes. I took several flash pictures, but the film lay undeveloped until several months later. Then I was disappointed and puzzled, for the tentacles had photographed a muddy brown. Thinking back, I deduced what had happened.

Reacting to the bright blue daylight the anemones had fluoresced, and to my eye, they had looked bright red. But when my flash fired, the reddish artificial light did not excite fluorescence, and the film recorded only the brown surface coloration.

If I had had my wits about me, I would have realized that normally one cannot see reds at 60 feet. What I was looking at is now a well-known phenomenon, but apparently the NATIONAL GEOGRAPHIC article of February 1956 was the first published account of underwater observation and photography of submarine biofluorescence.

Anyone who has ever thrust his mask-enclosed face under water knows all divers have big feet. And big hands, too. That is, they look bigger when seen through a diving mask. Because of the difference in refraction of light rays passing from water to the air inside the diver's viewing mask, everything underwater appears one-third larger than life.

Things look bigger and closer to the camera lens as well, so to restore objects to normal size and to maintain a full field of view, correcting lenses have been designed to replace the flat glass portholes of most camera housings. The correction can be made by curved glass, which all light rays strike at approximately a right angle, causing no distortion; a better corrector may be made of two lenses in combination.

Flipped onto its back (upper left), a blue starfish rights itself by turning a slow somersault on the seabed off Queensland, Australia. Mating cuttlefish entwine tentacles in a rare underwater photograph near New Caledonia in the South Pacific. Related to the squid and octopus, a cuttlefish camouflages itself by changing color in two-thirds of a second, or screens itself with a brown-black ink used by artists for pigment.

DOUGLAS FAULKNER. STARFISH 1/6 LIFE-SIZE; CUTTLEFISH LIFE-SIZE

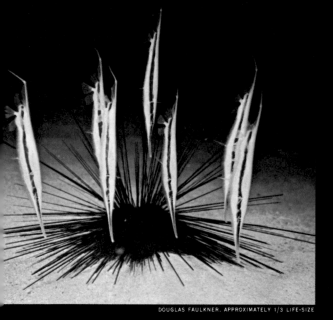

Bizarre patterns of the sea: Tawny blotches of a Sargassum fish (above) merge with weeds in a floating meadow of the Sargasso Sea. Featherlike shrimpfish swim head-down among prickly spines of a black sea urchin. Purplemouth moray eels, a menace to divers, bare needlelike teeth, and a translucent jellyfish trails delicate but deadly stinging tentacles. Bunched like a cluster of matches, nubby tentacles of a sea anemone shelter a spidery cleaner shrimp. A gaudy sea slug inches above coral polyps, and a silvery seahorse roams reef coral. The Nassau grouper scowls like a bulldog and, when hooked, fights like one.

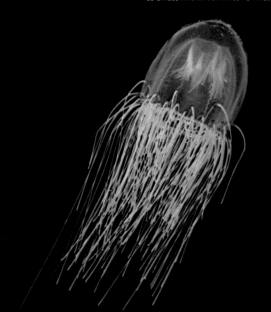

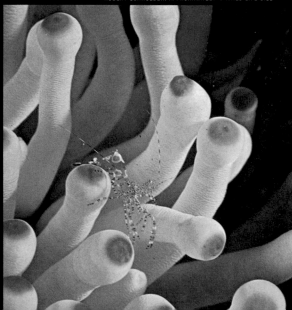

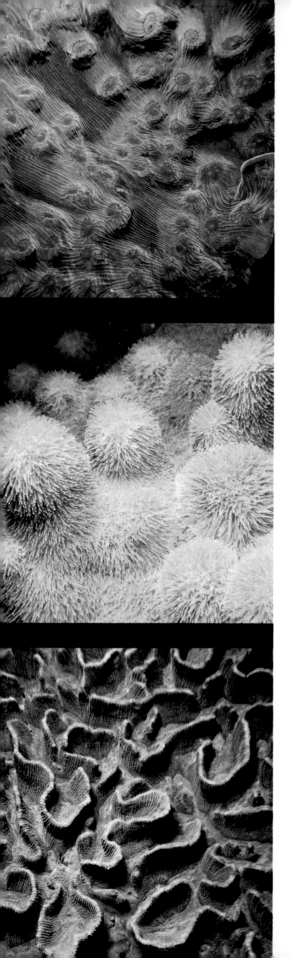

Some panoramic view masks with a curved plate give a wide-screen effect. I once tried a flat plate with slanting sides that gave me the impression I was swimming continuously in a steep-sided trench.

The average diving mask makes the diver feel like a horse in blinders; he can see straight ahead, but has no vision at the sides. Sometimes this can be frightening.

Once, in the Indian Ocean, as I bent over to focus my camera, something whipped over my left shoulder and hit the flash reflector with a resounding clang. I nearly spat out my mouthpiece in fright, thinking it might be a shark or barracuda. When the fish turned and came toward me, I recognized it as a big, harmless jack. The polished parabolic reflector must have attracted the fish by sending moons of light glancing through the water, like the wobbling chromed spoon of a trolling fisherman.

One of the underwater photographer's most valuable tools is the electronic flash invented by a witty and brilliant man named Harold E. Edgerton, now Professor Emeritus of Electrical Measurements at the Massachusetts Institute of Technology. In 1931 Dr. Edgerton, seeking a method of freezing the motion of high-speed machinery, built a flashing stroboscopic lamp bright enough to photograph by, and the high-speed electronic flash was born.

Through development and refinement, the instrument lost weight and its flash increased in brilliance. The flash—as brief as 1/100,000 of a second or less—could stop such high-speed motion as the wings of a bird in flight or the swing of a golf club.

Hunting algae and small invertebrates, orange-spot filefish skim past a cluster of staghorn coral in the Indian Ocean. The tubelike mouth of the three-inch fish contains incisor teeth that can nip off hard-shelled barnacles. Soft, pastel coral polyps build limestone homes throughout earth's warm seas. Fluorescing coral (top) reveals tiny rivulets of color. Cactuslike pillar coral (center) grows in spiny clumps, and maze coral (bottom) forms a network of twisting corridors.

DOUGLAS FAULKNER. APPROXIMATE SIZES: FILEFISH LIFE-SIZE; FLUORESCING CORAL 1/2 LIFE-SIZE; PILLAR CORAL 1/2 LIFE-SIZE; MAZE CORAL 1/4 LIFE-SIZE

109

Curiosity compels a sleek sea lion to linger for photographers working near Port Arthur, Australia. Another of the animals (below, right) *reveals a streamlined form ideal for underwater travel. An ungainly ten-foot-long manatee (below) lolls contentedly at the surface of a*

RON AND VALERIE TAYLOR (ABOVE); BILL DE COURT (LOWER RIGHT); AND JAMES A. SUGAR

Florida river while snorkelers scratch its chest and head. When alarmed, manatees can swim nearly 20 miles an hour, propelled by tail strokes.

Dimitri Rebikoff, a French engineer, adapted an electronic flash of his design to underwater use in 1949. He enclosed a lamp and mechanism in a long plexiglass tube filled with clear oil to withstand pressure that otherwise might crush the tube. With this instrument, the submerged cameraman was free of the fragile and bothersome flash bulbs that not only tended to bite the hand that held them but also flew upward to the surface if they escaped from the net bag.

In 1951 I went to M.I.T. to confer with Dr. Edgerton about his high-speed photographs of hummingbirds. Somehow we began to talk of my long love affair with the undersea world, and Dr. Edgerton said, "For a long time I have had an idea for an unmanned deep-sea camera knocking around in my head." He described an automatic camera synchronized with a flashing strobe light that would take 500 pictures at a charge.

On my return to Washington, I told Melville Bell Grosvenor, of the National Geographic Society, about the professor's idea. The Society's Research Committee subsequently furnished Dr. Edgerton with support to build and to experiment with the deep-sea automatic camera.

When the abyssal camera was ready, Dr. Grosvenor introduced Professor Edgerton to Captain Cousteau. The ebullient crew of *Calypso* quickly dubbed the inventor of the strobe light "Papa Flash."

At first the professor lowered the camera and a cylinder holding the flash mechanism on deep-sea trawling wires, but the wires were heavy and tended to break.

One wire snapped while a camera hung some three miles down on the north wall of the Puerto Rico Trench. Dr. Edgerton carefully noted the latitude and longitude, and the following year went back to look for it.

"We welded together all the scrap angle iron we could find on the ship," he recalls, "to make a snagging grapnel. Three miles of wire had gone down with the camera and I thought we might hook the tangle. We dragged the area all night, and then reeled in the wire. It took an hour to bring it up, and all we found wedged in the grapnel was one piece of coal. But my name and address are clearly marked on the camera, so, if anyone finds it, will he please send it back to me?"

The weight of three miles of wire alone can be enough to cause it to snap, not to mention the added drag of the equipment on the end of it. To circumvent this Dr. Edgerton came up with the idea of using a line of nylon, which has the same specific gravity as water and therefore would have no weight when submerged. It worked perfectly. In fact, when a camera caught on the bottom several thousand feet down, the highly elastic line actually pulled the 360-ton *Calypso* backward.

All the ship's company helped as Cousteau and Dr. Edgerton built an ingenious device in 1953. They turned a metal ladder into "a purely impromptu invention," the sea sled. The ladder's handrails served as runners when the ship towed it along the sea floor, and the camera strapped to its rungs ultimately took thousands of pictures of the bottom and its creatures.

Techniques and equipment have improved since the early experiments. Nowadays an echo sounder measures the distance to the bottom, enabling the surface operators to place a camera precisely with relation to the seabed. But "Papa Flash" is still dissatisfied with the area of sea bottom covered by his cameras.

"Under the sea," he says, "the weather is always bad for photography. At the very best, with super-clear conditions, an underwater camera seldom can be used even at 100 feet from the subject or the bottom. The light-scattering and absorption effects of the water act as fog does over land.

"A mapping aircraft can cover an area of 20 square miles from an altitude of 30,000 feet. Until recently, most deep-ocean pho-

Creeping in bottom ooze more than a mile down off Cape Cod, a sea spider, lighted by an electronic flas

tography was being done ten feet from the bottom, covering about 50 square feet at each exposure. If we could raise the cameras and take pictures from 30 feet, we could then take in 500 square feet of sea floor at a time."

Even at this rate, to photograph all of the continental shelves would take more than 640 billion exposures — or the equivalent of one exposure every 10 seconds by each of 10,000 cameras operating 12 hours a day for 40 years! Impossible as the undertaking now seems, the development of effective, large-scale underwater mapping techniques is an essential step toward fully utilizing the shelf zones for submarine

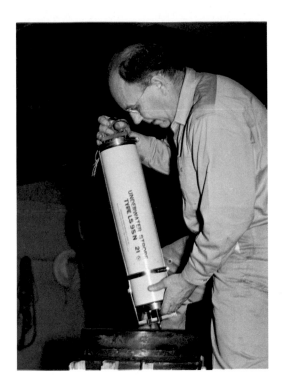

Pioneer of deep-sea lights and cameras, Dr. Harold E. Edgerton of the Massachusetts Institute of Technology lowers an early model strobe into a "pressure pot" for tests at simulated depth.

...sts a shadow over a five-pointed brittle star; a deep-sea fish, propelled by its whiplike tail, forages for food.

agriculture, fish farming, and mining.

Dimitri Rebikoff has designed Pegasus, a fascinating nine-foot-long torpedolike gadget powered by an electric motor. It "flies" like a miniature aircraft, whether operated by a diver stretched out atop its hull or remotely controlled. Cameras on Pegasus open their shutters in synchronization with a flashing strobe light to photograph strips of sea bed.

These applications of the unmanned abyssal camera are of immense importance, but as a working journalist, my first love goes to the hand-held cameras that I can use to make photographs of the strange beauty of the drowned world. Such equipment has come a long way since my experiments in the 1940's.

Dr. Edgerton and others have designed powerful hand strobe units that either work directly in synchronization with the camera or fire remotely, triggered by a photoelectric cell. A Belgian engineer, M. Jean de Wouters, has developed the world's first amphibious camera, the Nikonos. This small 35-mm camera needs no special housing because it is itself waterproof, like a diver's watch. And a remarkable new lens with a 94° field of view permits clearly focused exposures of large underwater objects at very close range.

The ocean deeps are the least-known parts of the earth, and creatures undreamed of in our zoology may well dwell there. Only recently have scientists given some measure of serious attention to "sea serpents." Of the hundreds of sightings over the centuries, many were too well-documented to be dismissed.

In 1930 Danish scientist Anton Bruun, trawling from the ship *Dana* off the Cape of Good Hope, brought up an astounding creature, long, flat, and translucent, with a diminutive head. It was a leptocephalus, the larva of an eel. In this stage most eels reach no more than four inches in length. But the leptocephalus hauled up in *Dana's* net was *six feet* long. If the larva continued to grow at the same rate as the common eel, the adult creature's length might have been as much as 70 feet.

Is this the Great Sea Serpent of yarns told by seamen down the centuries? I think it is, and one day an Edgerton camera, flashing in the abyssal blackness every 15 seconds, may well give us a portrait, beyond doubt or cavil at last, of the most storied sea creature of all.

Sea sled improvised by Cousteau (in blue shirt) and his crew swings aboard Calypso, *anchored in Nice Harbor, France. Forerunner of more sophisticated models, the sled supported an Edgerton camera and strobe synchronized to click every 15 seconds while being towed along the bottom. Holes drilled in the tubular frame admitted seawater to equalize the deep pressures. Bumping along at 10,000 feet, the apparatus lighted a lava-studded valley (opposite) of the Mid-Atlantic Ridge.*

6

The Sea's Dark Museum

BY GEORGE F. BASS

Bronze Poseidon, or possibly Zeus, bears the patina of 2,300 years' immersion in the Aegean Sea. Divers in 1928 brought up the Greek masterwork off Cape Artemísion. It now stands in the National Archaeological Museum, Athens.

IN THE TWILIGHT of the Bronze Age, a tiny merchant ship set sail one day from the Syrian coast and headed westward across the Mediterranean. Troy had fallen a few years before. The great Mycenean civilization of Agamemnon and Achilles lay dying. The year was about 1200 B.C.

The little freighter put in at the island of Cyprus to take on her main cargo. Stevedores wrapped matting around four-handled copper ingots, weighing nearly 50 pounds apiece, and stowed them on a cushion of brushwood in the vessel's hull. They lugged aboard smaller ingots of tin and bronze, along with wicker baskets of broken tools and weapons and other scrap metal. Hugging the coast of Turkey, the ship sailed on toward the Aegean Sea.

Near the stern, in an area lit at night by a single oil lamp, a tinker sat amid his tools. He must have been content. Everything he needed for a successful enterprise was at hand: materials for casting, hammering, sharpening, and polishing bronze implements; sets of weights for use in trading along the route; even a jar full of glass beads and a few unworked lumps of crystal for minor bartering.

For centuries, ships carrying copper from the mines of Cyprus followed this route. The number of Bronze Age ingots brought up by fishermen and sponge divers along the Turkish coast, however, testifies that storms and shoals took a heavy toll.

Our ship was one of the unlucky ones. Rounding Cape Gelidonya near the southwest tip of Turkey, she sank, her hull coming to rest on the rocky, nearly barren bottom 90 feet below. With no sand or mud to cover and protect them, most of the wooden remains were soon devoured by shipworms. Currents swept away pottery and light cargo. But a ton of metal cargo lay undisturbed for the next 32 centuries.

The day I first heard of Cape Gelidonya was the day I first heard of Peter Throckmorton. The two names would change my life. It was the fall of 1959. Dr. Rodney S. Young, chairman of the Department of Classical Archaeology at the University of Pennsylvania, called me to his office. I had

117

THE AUTHOR: *In 1961, while a doctoral candidate at the University of Pennsylvania, George F. Bass led history's first methodical underwater archeological dig, at Gelidonya, Turkey. Contributor to numerous periodicals and author of* Archeology Under Water, *Bass now lives in Cyprus. He is president of the American Institute of Nautical Archeology.*

only recently arrived as a graduate student after two years in the U. S. Army, but several years before I had worked for Dr. Young as a student assistant on an excavation in central Turkey.

"I've just received a letter from John Huston of the Council of Underwater Archaeology," he said. "A journalist named Peter Throckmorton has found a Bronze Age shipwreck and wants to know if the University Museum will sponsor its excavation. Want to learn how to dive?"

I accepted immediately. Not only did I hope to specialize in the Bronze Age—between 3000 and 1000 B.C.—but I had grown up reading everything I could find on diving and salvage. Yet I hadn't had— or made—the opportunity to dive myself.

Peter and I met a few months later, on a rainy December night in New York. In his apartment we talked until dawn as we leafed through his astonishing log of two summers spent with Turkish sponge divers.

"I am convinced," Peter said, "that we can excavate a sunken ship as carefully and accurately as archeologists unearth tombs and temples on land."

We had little money, equipment, or experience. Half of the staff of eight that finally set out had never dived before. Yet we were determined to prove Peter right.

By summer we had set up camp in Turkey, on the beach of a cliff-ringed bay an hour's run from the inhospitable rocks of Cape Gelidonya. Each morning we sailed to the cape in two sponge boats.

Mapping our Bronze Age wreck presented the first problem. Any diver can salvage antiquities; an archeologist must keep careful records, so he and others can later interpret his work. But a limestone concretion half a foot thick encased most of the wreck's metal cargo. Limited in the time we could spend underwater, we knew no safe way to extract hundreds of delicate, camouflaged objects from their cocoons. Chief diver Frédéric Dumas solved our problem. "It would be easier to raise big hunks of boulders and let topside people take them apart," he suggested.

So for weeks we chiseled around great clumps weighing up to 300 pounds. We even used an automobile jack to break pieces apart. A diver would attach a broken-off lump to a lifting balloon, inflate the balloon with his air hose, and guide his trophy toward the sponge boat on the surface.

Beneath a corner of one mass we discovered an ancient merchant's seal, used to stamp receipts on clay tablets. Religious symbols inscribed on it told us the cylinder was made somewhere in North Syria centuries before its fateful voyage; by 1200 B.C. it was already an heirloom.

In the last days of our campaign the weather turned bad, and we fled exhausted. But the work had scarcely begun; it would take years of research to fit our wreck into its historical context. We had opened the door to a new understanding of Semitic seafarers who plied the Aegean centuries before their famed descendants, the Phoenicians. But in the summer of 1960 we knew only that we had accomplished on the sea bed what Dumas later described as "the first methodical excavation carried to completion" beneath the sea.

Dumas spoke with authority. Our success was firmly based on the previous decade of underwater archeological research by the French pioneers of Aqua-Lung diving: Jacques-Yves Cousteau, Philippe Tailliez, and Dumas himself. As early as 1948, the French explorers had tried out the new equipment during a reconnaissance dive to the famous Mahdia wreck off Tunisia.

Scuba diver 120 feet down inspects a jumble of amphorae, or wine jars, spilled from a Greek merchantman that sank about 230 B.C. near Massalia (modern Marseille). In 1952 an archeological expedition began the salvage of some 3,000 amphorae from the vessel's cargo.

This ancient ship met her fate in the first half of the first century B.C., going down with a hold full of bronze and marble statuary and more than 60 marble columns. Salvage operations from 1908 to 1913 retrieved enough to fill several galleries of a Tunisian museum, but the suited, helmeted divers of that era lacked the mobility and dexterity to do delicate excavating.

"We were merely scratching at history's door," Cousteau wrote. But the 1948 dive marked an important beginning. Four years later Cousteau undertook the first large-scale marine archeological excavation to use modern diving equipment.

At Grand Congloué Island, ten miles east of Marseille, a Greek wine-carrier had sunk in the late third or early second century B.C. In a project that took five years and one diver's life, Cousteau proved a new method of excavating underwater with a suction hose. Although the excavation was never completed, the thousands of pieces of pottery brought to the surface shed considerable light on ancient times, and sections of hull told of sophisticated Greek shipbuilding techniques.

Cape Gelidonya. Mahdia. Grand Congloué. Three tiny dots on the map: three ancient shipwrecks. How many more such dots should an exhaustive map show?

If only one ship had sunk each year since the first seafarers crossed the Aegean, there would be ten thousand wrecks awaiting exploration. But many more than one a year have sunk and lie submerged: fishing boats, merchantmen, warships, pleasure boats, ferries. Each carries clues to ancient civilizations, migration patterns, commerce, or naval warfare.

Substituting air for brawn, divers recover ages-old cargo near Turkey. The balloon lifts a sea-welded mass of copper ingots and bronze tools recovered by archeologists from a Bronze Age trading vessel that sank in 90 feet of water off Cape Gelidonya 32 centuries ago. Buoyed by air from a diver's Aqua-Lung, the amphora (below), part of a wine shipment lost near Yassi Island, will rocket to the surface when released.

Bronze Age relics lie in sunlight after three millenniums in the sea. Artifacts from the

For years divers have picked over sunken wrecks in the Mediterranean, finding worth only in sponges and corals and leaving the artifacts to gather bright crusts of sea life. We archeologists sadly imagine the many vanished remnants of ancient cargoes that fishing nets must have caught, particularly classical sculptures. Bronze figures had scrap value and were melted down in furnaces. Marble statuary was burned for lime or piled on breakwaters.

At painfully long intervals a few sculptures found their way into museums. On a day in 1900, two Greek sponge boats, threatened by a spring gale, sought shelter near Andikíthira Island northwest of Crete. A diver, Elias Stadiatis, hunted sponges in the unfamiliar waters. Suddenly he stared in awe at an eerie array of bronze and marble figures of men and horses. To take tangible proof of this ghostly scene to his captain, Demetrios Kondos, he clutched a large bronze arm and ascended the 170 feet to the surface.

His discovery spurred a government-sponsored salvage effort—the first attempt

PETER THROCKMORTON, NANCY PALMER, INC.

Cape Gelidonya wreck include bronze picks, chisels, axes, and oxhide-shaped copper ingots. Expedition director George Bass and his wife,

Ann, apply plastic preservatives to bits of wood from the ship. Mrs. Bass delicately brushes the largest piece of recovered timber.

to excavate a shipwreck for the benefit of archeology. Scholars believe the wreck was a Roman trading ship, on its way home from the eastern Mediterranean in the first century B.C. Tableware and storage jars helped fix the date of the wreck; so did a bronze astronomical device. Its discovery altered our concept of the technological achievements of early Mediterranean scientists. In 1959, Dr. Derek J. de Solla Price, Professor of the History of Science at Yale, announced that the device with its gears and dials probably served as a calendar to calculate the motions of stars and planets.

Sculptures rescued by this early expedition found a place of honor in the National Archaeological Museum in Athens. The museum's prize, the heroic Poseidon (or possibly Zeus), was recovered by divers near Cape Artemísion in 1928.

The Piombino Apollo, pride of the Louvre, was netted by fishermen off the coast of Italy in the early 19th century. "The sea offers our one hope of finding original bronzes by the great Greek sculptors," points out Dr. Brunilde Ridgway,

expert on classical sculpture at Bryn Mawr. Most of the Greek bronze masterpieces were melted down or hammered into scrap, and are known only from descriptions or from later Roman copies in stone.

By the time divers reclaimed the treasure near Andikíthira, evidence of another venturesome race of seafarers had come to light far to the north. But these finds were not true wrecks. Remains of the strong, graceful ships of the Vikings were first uncovered on land, buried with Norse nobles of a thousand years before. Best known is the Oseberg ship, which sailed from about A.D. 800 to 850. Unearthed in 1904 and restored, it displays its sweeping lines at the Viking Ship Hall in Oslo.

By the 1950's Scandinavians had progressed from "dry" to "wet" archeology and were searching out Viking shipwrecks in shallow offshore waters. When two amateur divers reported a hulk lying submerged in Roskilde Fjord in easternmost Denmark, investigating scientists dis-

covered not one wreck but five, each of a
different type. About A.D. 1000 someone
had filled them with stones and sunk them
to block the approach to Roskilde.

For my friend Ole Crumlin-Pedersen,
who excavated the ships with Olaf Olsen,
the wrecks lay not too deep but too shallow! "You can stand up in the water," Ole
told me. To solve the unusual problem,
Ole and Olaf built a cofferdam around the
site, then drained out the water. Like jack-
eted sloths the excavators worked, arms
and heads hanging down from wooden cat-
walks laid across the ooze. They cleaned
each wood fragment with a gentle jet of
water and mapped its location. Now reconstructed, the five hulls attract thousands
to the Viking Ship Museum at Roskilde.

Myriad other wrecks, extraordinarily
well preserved, lie in cold northern waters.
The proof was there for all Europe to see
on television when in 1961 the warship
Vasa broke the surface of Stockholm's harbor. Built to be the pride of the Swedish
fleet, she sank on her maiden voyage in
1628—unable even to reach the open sea.
Her raising was a triumph of persistence
for a young petroleum engineer, Anders

*Straddling scaffolding 100 feet down, a diver
photographs timbers of a Byzantine merchantman wrecked near Yassi Island 1,300 years ago.
Iron grids enable archeologists to plot the vessel's
size and type. For centuries, heavily laden trading ships swarmed among the islands off the
Turkish coast, and Yassi's treacherous reef
claimed at least a dozen of them.*

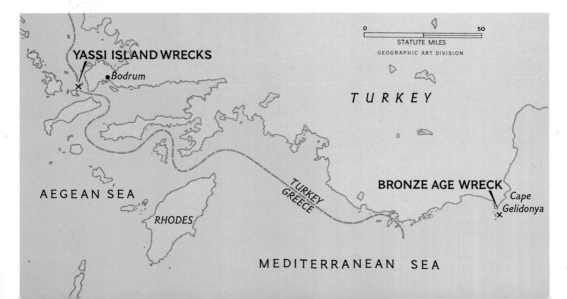

Franzén, from the Swedish Admiralty.

Vasa's recovery was no accident, but the result of determination born of a single, basic idea. "Sunken wooden vessels are destroyed by shipworms," Anders explained when we traded ideas at a diving conference. "And I knew there were no shipworms in the Baltic Sea. It's not salty enough." He reasoned that *Vasa*, if he could locate her, would show little deterioration.

For two years he dragged the harbor with grapnels. Then he noticed a suspicious hump on contour maps. He invented a tiny core sampler and dropped it to the hump. Up came a plug of oak. *Vasa* was his!

In cold, inky darkness 110 feet down, helmeted divers burrowed six tunnels in the mud beneath *Vasa;* through these they threaded steel cables which were then secured to salvage pontoons on the surface.

ROBERT GOODMAN (UPPER LEFT); NATIONAL GEOGRAPHIC PHOTOGRAPHER BATES LITTLEHALES (ABOVE); AND THOMAS J. ABERCROMBIE, NATIONAL GEOGRAPHIC STAFF

Research vessel for underwater archeologists,
Asherah—named for the Phoenician sea god-
dess—prowls for sunken ships off Turkey. The
two-man submarine reaches depths of 600 feet.
Built for the University of Pennsylvania with
grants from the National Geographic Society
and the National Science Foundation, the sub
can map a shipwreck site in hours—a task that
takes divers many weeks. An undersea artist
with graphite crayon and a sheet of frosted plas-
tic (opposite) plots wine jars at the Yassi Island
wreck, before the building of Asherah. *Arche-*
ologists topside, using divers' drawings and
photographs, chart the cargo. Relics tagged for
identification head for the surface (right).

Bronze bar, chain, and hooks from the Yassi Island wreck form a steelyard, used by Byzantine traders for weighing cargo. The scale, almost identical to those seen throughout Turkey today, carried a lead-filled counterweight — a ten-inch bust of the goddess Athena.

Still intact after 13 centuries on the ocean floor, an earthenware plate, ribbed cup, and wine pitcher served the captain at his table. Resin lining of the jar prevented it from sweating. Clay "wine thief," a type of pipette, drew wine from amphorae. Cooking pot (far left) lay in the galley area. Gold coins of the Byzantine Emperor Heraclius, who reigned A.D. 610-641, indicate the ship sank during the first half of the seventh century.

THOMAS J. ABERCROMBIE, NATIONAL GEOGRAPHIC
STAFF (COINS), AND ROBERT GOODMAN

Eighteen times the pontoons were partially filled with water to lower them; then, after the cables had been tightened, the pontoons were pumped out so they floated upward, raising *Vasa* slightly. Each time the warship was shifted to somewhat higher ground on the harbor bottom. Finally, in a last effort aided by hydraulic jacks, *Vasa* reached the surface. Today, in Stockholm's Vasa Museum, the ill-fated man-of-war belatedly claims her proper glory.

Forty years before *Vasa* went down, an entire war fleet suffered losses on such a staggering scale its name became synonymous with defeat at sea: the Spanish Armada. In 1588 its 130 ships carrying some 30,000 men sailed from Spain for England. But victory eluded them, and scarcely half the Armada returned home.

Of the ships that sought to escape the pursuing English by circling Scotland, four have been discovered in recent years, two off the coast of Ireland. *Santa Maria de la Rosa* has been partially excavated by a British team under Sydney Wignall; the galleass *Girona* yielded her secrets to Belgian diver Robert Sténuit.

On the night of October 26, 1588, the proud *Girona* was carrying 1,300 souls, including Spaniards rescued from previous disasters. A raging storm drove her onto the rocks near the Giant's Causeway in northern Ireland.

In 1968 and 1969, Sténuit's divers battled chilling temperatures, writhing kelp, and the surge of surf to harvest a rich reward. From the watery grave they plucked cannons, navigational instruments, garments, galley wares, silverware, gold and silver coins, and jewelry. At a depth of 30

Roped to a salvage ship, a 6½-foot cannon leaves its undersea grave — the wreck of a 16th-century European cargo ship 40 feet down off Bermuda. Salvagers armed with a sand-clearing vacuum (right center) search the hulk. Other Bermuda finds: the paddle wheel of the Mary Celestia, *a Confederate gunrunner sunk in 1864; and a 2¼-inch emerald-studded gold cross — valued at more than $100,000 — found near the Spanish* San Pedro, *sunk in 1594.*

feet, Robert came upon a thin gold ring.

"I carried it up out of the sea which had held it so long," he wrote in NATIONAL GEOGRAPHIC. "Safe aboard our boat the ring lay cold and wet in my black-gloved hand, glowing softly under the pale Irish sun. Carved upon it was a tiny hand, offering a heart, and these words: '*No tengo mas que darte* — I have nothing more to give thee.' To me that ring is the most beautiful, the most touching treasure of the Armada."

In the Western Hemisphere, divers bringing up cannon, coins, and jeweled ornaments from Spanish treasure ships have led archeologists and historians to many important sites.

Shallow waters of the Florida Keys, the Florida Straits, the Bahamas, and Bermuda are littered with storm-wrecked ships that, for three centuries after 1500, hauled gold and silver from Mexican and Colombian mines. Commercial diver Arthur McKee pioneered in locating and excavating treasure-rich wrecks. In 1951, he accompanied engineer-inventor Edwin A. Link and Mendel L. Peterson, then the Smithsonian Institution's diving Curator of Naval History, on a study of a sunken vessel near Marathon Key. With an air lift they recovered much material from the *Looe*, a mid-18th-century British frigate.

From a 1733 Spanish silver-fleet wreck

Relics of a defeated navy: Gold and silver coins uncovered in the wreck of the Girona, *a Spanish Armada galleass sunk in 1588 off the coast of Northern Ireland, intrigue archeologists Robert Sténuit, right, and Francis Dumont. Often using a homemade metal detector (above) to search the murky, turbulent waters, the divers retrieved a treasure of precious and everyday items, including gold-framed cameos of lapis lazuli — the jewelry of noblemen — and an astrolabe like one raised from another wreck (below).*

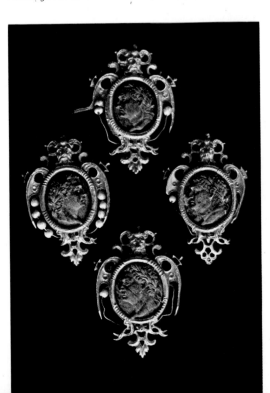

NATIONAL GEOGRAPHIC PHOTOGRAPHER JOSEPH J. SCHERSCHEL

that McKee had been working for some years, the trio brought up in 1953 and 1954 a trove of historical material. "This wreck was one of the few cases in these shallow waters," Peterson observed, "where the entire bottom of the reef-torn ship remained intact—because it was buried in sand and under 28 feet of water, out of the surging of the waves. Shallow coastal waters are a help to the diver, but ships wrecked there take a thrashing from the sea that sometimes strews cargo over miles of bottom. That leaves the archeologists plenty of important but jumbled artifacts—quite different from the relatively undisturbed wrecks in deep Mediterranean waters."

I have made only one dive in North American waters, and that in 20 feet off Cape Fear, North Carolina. I went down to view the Civil War blockade runner *Ella* in an area where archeologists periodically try, but not very successfully, to find famous historical vessels. The surge of the waves buffeted me, and currents swept me against the iron plates of the paddlewheeler. I was amazed that anything could be accomplished under such difficult conditions.

Spectacular finds of gold and jewels brought up in 1955 by Edward (Teddy) Tucker from a 1596 Spanish wreck in the clear reef waters off Bermuda quickly converted dozens of divers into treasure hunters. But archeologists and historians were excited by the ship's old weapons, utensils, and instruments, and by very rare artifacts, including a staff-of-office, made by Carib Indians. Beginning in 1965, Tucker became an important member of a

Archeologist-diver Peter Throckmorton (second from left) studies with Italian salvagemen a diagram of a Roman shipwreck he discovered in the Gulf of Taranto. The vessel, carrying marble coffins from Turkish quarries to Rome, went down in a violent storm during the second century. Protected from rotting by the sarcophagi, the wooden hull and planking revealed ancient methods of ship construction. William Phelps, the expedition's associate director, examines bronze spikes used to fasten planks to the hull frame.

Smithsonian diving team that has carried out archeological investigations.

To teach divers archeological techniques and to control the search for and salvage of wrecks in Mexican waters, Pablo Bush Romero organized the Club of Exploration and Water Sports of Mexico (CEDAM) in 1959. Club divers in 1967 discovered in Yucatan waters one of the oldest ships yet found, an early 16th-century Spanish vessel.

AT PORT ROYAL, Jamaica, hundreds of acres of the 17th-century pirate town lie under water into which it slid during the great 1692 earthquake. Having had a look at it in 1956 with McKee and Peterson, Edwin Link returned in 1959 with his new *Sea Diver II,* specially equipped with bottom-sounding and metal-detecting devices and heavy-duty lifting gear. Link, his wife, Marion, Peterson, and a six-man team of Navy divers recovered hundreds of artifacts. Later, diver Robert Marx continued for two years to bring up thousands more.

With all its disappointments, expense, and hazards, diving for treasure looked more enticing than ever after Kip Wagner's success. Wagner, a former housebuilder, had turned coins over with his toes in the sand north of Vero Beach, Florida. He pursued their source relentlessly and determined it was a Spanish treasure fleet wrecked in a 1715 hurricane. With persistence and luck, Wagner located eight ships of the 12 that were lost. By 1970, the coins, jewelry, and bars of precious metals he recovered totaled more than $6,500,000 in antique value.

The salvage license granted by Florida to the company Wagner formed—the Real 8 Co., Inc.—permitted him and his ten partners to keep 75 percent of the fortune. Yet Wagner wrote in 1965: "When I look back on our struggles over the years, the money value of over a million dollars seems almost meaningless. The real treasure lies in our having touched hands with history."

A cynic might not believe those words, but though I never met the late Kip Wagner, I'm sure he meant them. For I, too, know there are rewards as valuable as gold: the romance of exploration, the thrill of discovery, the bonds of friendship forged under adversity. Each diver takes his personal motives with him beneath the waves.

Marine archeologists working in the Mediterranean can keep little of what they find. Timbers and cargoes of ancient hulks usually belong to the country in whose waters they lie. Yet, like the astronomer who photographs, maps, and analyzes the stars but cannot own them, we are content to work for the knowledge these lost argosies hold.

Each shipwreck is a miniature Pompeii. Its dissection and study can produce a vivid picture of a lost moment in time. We uncovered two such storytelling wrecks in the graveyard of ships Peter Throckmorton found near Yassi Ada, Turkey. Peter by then had moved on to locate and excavate other sites in Greece and Italy.

Our team began its investigation of the first of the Yassi Ada wrecks in 1961 for the University of Pennsylvania Museum. My first task was to improve the techniques used on the Bronze Age wreck at Cape Gelidonya the year before. Our new find, a Byzantine wine carrier, lay buried in sand and mud, and we hoped to find much of the hull still intact. We planned to piece together a blueprint of the vessel, plotting every fragment of wood we discovered, every nail, every bolt hole. To do this we erected a scaffolding over the entire site. Moving a 12-foot-high camera tower across this grid, we systematically recorded the excavation. Later we invented methods of mapping underwater, using stereophotography, an aerial mapping technique.

For four consecutive summers we worked, spending 1,200 man-hours at a depth of 120 feet. After removing the ship's cargo of nearly a thousand empty wine jars, sending many to the surface in baskets lifted by air-filled balloons, we carefully dug into the sand below by sweeping it into the open mouth of a suction hose.

Slowly we filled in the picture of a 65-foot, square-rigged merchantman, built in the first half of the seventh century. We have learned how much the ship cost its

Drowned hulk of a Greek merchantman rests 90 feet down in the Mediterranean off the port of Kyrenia, Cyprus; grid pipes divide the wreck and surrounding sand into working areas for a team of diving scientists. Bubbles from their tanks rise toward the surface. During the summers of 1968 and 1969, archeologists directed

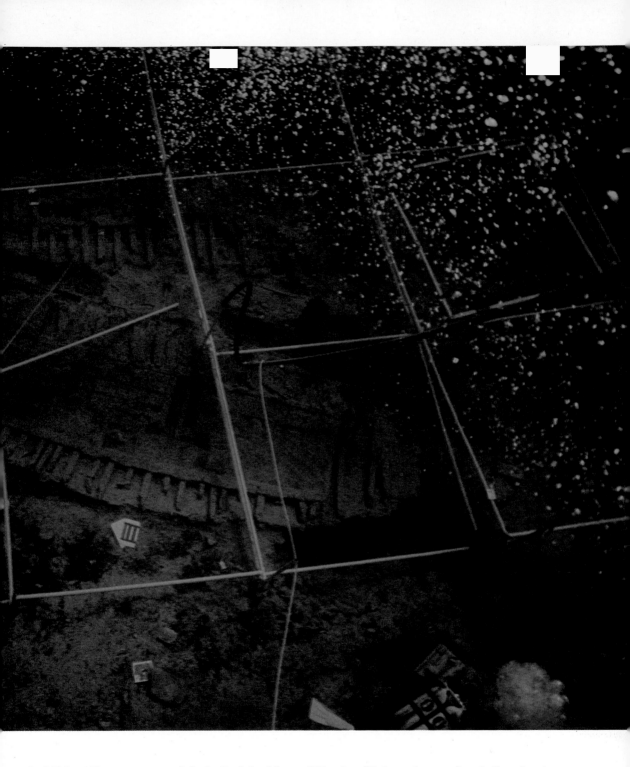

by Michael Katzev excavated the hull of the 50-foot cargo ship, painstakingly raising it piece by piece. When reconstructed, the vessel will stand in Kyrenia's Crusader castle. Built in the early part of the fourth century B.C. and sunk about 80 years later, this oldest-known Greek ship bore a cargo of thousands of almonds and more than 400 wine-filled amphorae when it foundered—probably in seas stirred by a sudden storm. A lead casing one-eighth inch thick sheathed its pine hull as protection against shipworms; but apparently the lead only shielded damage from the sailors' eyes—teredo holes riddled the planking and hastened the craft's end.

137

Hurricane shatters a homebound Spanish treasure fleet off the east coast of Florida, July 31, 1715

owner, how much money the captain carried on board, what the crew ate, what tools the carpenter used, where the helmsman stood to heave on his steering oars, how the boatswain foraged for wood and water to supply the cook in his cramped, tile-roofed galley. We even learned the names of some of those who sailed on the final voyage, and the route that led to disaster on the Turkish reef in A.D. 625.

A number of archeology students who worked with me at Yassi Ada became proficient divers, and some have been expedition leaders. Already Michael and Susan Katzev have conducted an exemplary excavation of the oldest seagoing hull ever recovered—a small Greek coastal trader of about 300 B.C. The lead-sheathed hull, constructed with edge-joined planks, proved that Roman shipwrights borrowed

Waves swallowed $14,000,000 in gold and silver. Divers began work soon after, recovering half the loss.

directly from their Greek predecessors.

Not only shipwrecks beckon underwater archeologists. Entire cities have been swallowed by a seemingly insatiable sea.

Dr. Nicholas C. Flemming, an English geologist, swam through the watery streets of a Bronze Age town he discovered in shallow water in southern Greece. Earlier he had mapped the remains of submerged Appolonia off the coast of Libya. The Roman walls of Baia still stand beneath the Bay of Naples. I confidently expect that one day the classical Greek city of Helice, which slid into the Gulf of Corinth during an earthquake, will be located.

On my most recent excavation at Yassi Ada, we had relatively calm, clear water but faced an old problem: the time limitation on divers working at 140 feet. To go deeper and stay longer, we had been forced to employ new underwater tools.

In the undersea twilight, I "walked" on my fingertips across the hull of a sunken Roman ship, gently "dusting" away silt. Now I heard the whine of our most sophisticated piece of equipment, *Asherah*. This maneuverable two-man submarine, named for a Phoenician sea goddess, can dive to 600 feet, stay down for hours, and move at speeds up to four knots.

I ducked into our underwater "telephone booth," another innovation. Here, fully 140 feet beneath the surface, I removed both mask and mouthpiece; dry from the chest up, I breathed deep of the fresh air being pumped from topside into the transparent plastic dome as I watched the mapping operation going on outside.

Asherah pilot Yüksel Eğdemir steered a level course 20 feet above the site. Copilot Don Rosencrantz watched the scene on his closed-circuit television, pushing a button every few seconds to trigger two exterior cameras and strobe lights. Within minutes, *Asherah*'s cameras had recorded the entire site—the first undersea use of aerial survey techniques by a submarine.

Physicist Jeremy Green had picked up signals from metal objects surrounding the wreck with his metal detector, and marked them for examination by the next diving team. After telephoning our information to the diving barge anchored far above, Jeremy and I departed for the submersible decompression chamber.

Up the slope we swam toward Yassi Ada, following the white cord that led to the chamber, a bright yellow sphere of steel 6½ feet in diameter. Entering through the bottom hatch, we doffed our tanks and sat down to decompress—still underwater, but in dry comfort. It had been a good dive. We had accomplished more in the last half hour, I reflected, than in weeks at Cape Gelidonya seven years before.

As 20th-century technology gives us more tools like these, archeologists will reach beyond wave-battered offshore wrecks. At deeper sites, we may find sunken vessels still intact, nautical time capsules that could offer important new clues for our harvest of history.

Weathered steps of a treasure hunter's beach cabin hold a king's ransom in pieces of eight, gold ingots, chains, and doubloons, and delicate K'ang-hsi porcelain. In the early 1960's Floridian Kip Wagner launched a search that has recovered more than $6,500,000 in treasure from the ill-fated Spanish fleet. The gold doubloon below, actually about the size of a silver dollar, bears the name of King Philip V and the date 1714. At left, an elated diver surfaces with two of the gold coins.

NATIONAL GEOGRAPHIC PHOTOGRAPHERS BRUCE DALE (LEFT) AND ROBERT OAKES (BELOW); LUIS MARDEN, N.G.S. STAFF (OPPOSITE)

7

Taxis
to the Deep

BY R. FRANK BUSBY

Designed to descend 15,000 feet, the
Aluminaut *proved too expensive for most
ocean work and went into storage
in 1970. Demand for submersibles
by private industry has tended toward
simpler vehicles useful to 1,500 feet.*

WITH A LAST GLANCE at the darkening sky, I scrambled through the hatch of the gleaming submersible and made my way aft to the scientific compartment. Minutes later the hatch cover clanged shut and Capt. Don Kazimir spoke on the radio.

"*Privateer*, this is *Ben Franklin*. All checks are positive. We're ready to dive. Over."

"*Ben Franklin*, this is *Privateer*. Permission granted. Dive!"

It was July 14, 1969, and my five teammates and I were beginning an underwater journey that would carry us from West Palm Beach, Florida, to a point 300 miles south of Nova Scotia—1,534 miles—propelled only by the current of the Gulf Stream. A month would pass—30 days, 11 hours, and 58 seconds, to be exact—before we would reopen the hatch.

This journey was a major milestone for me. As a project leader for the U. S. Naval Oceanographic Office's Deep Vehicles Branch, I had been diving for five years in undersea craft similar to the 49-foot-long *Ben Franklin,* but generally smaller. To develop instruments and operating techniques for a proposed surveying submersible capable of descending 20,000 feet, my teammates and I had worked with five undersea vehicles to depths of 6,000 feet. Now we had our first opportunity to put all these instruments on one submersible and see how a prototype surveyor would perform on an extended mission.

British oceanographer Ken Haigh and I were the scientific crew of a unique survey vessel: *Ben Franklin*'s 29 viewports gave us a front-row vantage, while our instruments provided physical, chemical, biological, geological, and geophysical measurements of our surroundings.

The choice of the Gulf Stream had been based on several factors: It would carry us over various types of ocean bottom; its upper waters would keep us warm; and we already knew enough about this powerful "river" to anticipate where it might take us and what we might encounter.

As we made our initial 1,600-foot descent to the bottom of the Florida Straits, Jacques Piccard—tall, slender originator and leader of the expedition—came aft to inquire in a soft Swiss-French accent:

"How does it go, Frank?"

"Everything looks fine, Jacques. We're ready when you are."

Our other two teammates, chief pilot Erwin Aebersold and NASA's Chester May, were equally enthusiastic, and Jacques turned his attention to the underwater world outside. Having already plumbed some of the seas' greatest depths, he was about to realize still another ambition as the submersible he had designed for Grumman Aerospace Corporation began her Gulf Stream Drift Mission.

But long before the sophisticated *Ben Franklin*, other men had penetrated the ocean depths. Until the early 1900's, success was measured in a few hundred feet. Then in 1930 the curiosity of naturalist William Beebe and the talent of engineer Otis Barton combined to take the two men and their bathysphere to a depth of 1,426

feet, where they described a world of incredible beauty, hostility, and challenge. In 1932 they improved their record to 2,200 feet, and two years later to 3,028.

Nearly 15 years passed before anyone made a serious attack on the bathysphere's 1934 record. Then, in the late 1940's, a strange-looking "deep boat" or bathyscaph appeared in the Mediterranean. Its builder was Jacques Piccard's father, Professor Auguste Piccard. The Swiss physicist and aeronaut adapted certain principles of his stratospheric balloon FNRS in constructing the bathyscaph FNRS-2. A passenger sphere 6.7 feet in diameter, made of steel 4 to 6 inches thick, was attached beneath six encased cylinders of gasoline. Initially the lighter-than-water gasoline float and the heavier-than-water sphere and ballast were in equilibrium. If seawater was admitted to a regulating cylinder, the total weight increased and the sphere descended; if iron-shot ballast was jettisoned, the descent would be reversed and the gasoline-filled float would cause the sphere to rise toward the surface.

With this underwater "balloon" the innovative scientist and his fellow researchers began an assault on the depths that eventually overcame a seemingly endless array of problems. After 12 years, on January 23, 1960, Professor Piccard's son and U. S. Navy Lt. Don Walsh peered through the bathyscaph *Trieste*'s windows at the bottom of the Mariana Trench, 35,800 feet below the surface. Man had reached the ocean's greatest known depth.

With *Trieste*'s achievement, builders of future deep-sea vehicles were freed from a costly race for depth records and could concentrate instead on performance.

THE AUTHOR: *Since 1964 R. Frank Busby has been conducting undersea surveys for the U. S. Navy and evaluating submersibles as survey vehicles. He was Chief Scientist aboard* Ben Franklin *during its 1969 Gulf Stream Drift Mission, and has written numerous articles on oceanography and undersea surveying. He knows more about research submersibles, wrote Jacques Piccard, than any other living man.*

WILLIAM BEEBE

Clambering over protruding door bolts, naturalist William Beebe squeezes from the steel bathysphere that in 1934 carried him and designer Otis Barton (in shorts) to a record 3,028 feet off Bermuda. From the three-inch-thick quartz window of the "lonely sphere," Beebe said, he "peered into the abysmal darkness ... isolated as a lost planet in outermost space."

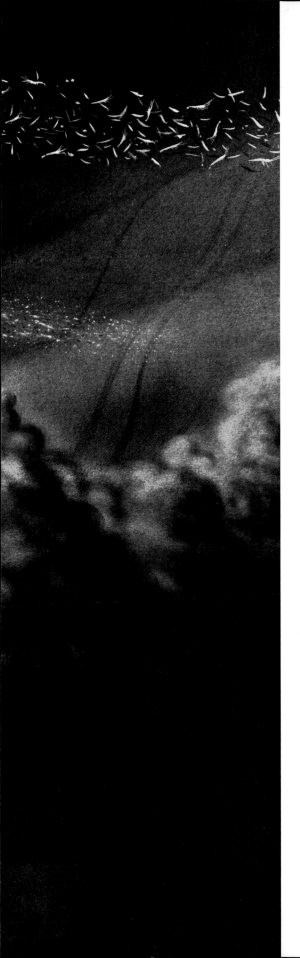

Ben Franklin was one of some 100 manned submersibles, progeny of the bathyscaph, that were constructed between 1960 and 1973. Larger but otherwise similar in many respects to her sister submersibles, the *Franklin* was steel-hulled and battery-powered, had acrylic plastic viewports, and—unlike the bathyscaph—was capable of comfortable, long-term cruising at depths to 2,000 feet.

The characteristics of mobility, maneuverability, and extended cruising capability, now common to most submersibles, were developed in the search for the nuclear submarine *Thresher*, which in 1963 carried 129 men to death in 8,490 feet of water. The search operations that located her pointed up the cumbersome bathyscaph's shortcomings, and clearly defined a need —with which oceanographers who used *Trieste* for research concurred—for smaller, lighter, more powerful and versatile craft.

The response was amazing. Prior to 1960, only three manned submersibles existed, including Jacques-Yves Cousteau's diving saucer. Less than ten years later there were 50 small, versatile submersibles, of varied design—and with names like *Alvin, Aluminaut, Deep Diver, Deepstar, Star III, Dowb, Pisces*—plunging through the once-impenetrable depths with impunity.

In early 1966, when an aircraft collision showered bombs and wreckage on the coast of Spain, the search for a lost hydrogen bomb indicated the rate of progress in the three years since the *Thresher* search. Hampered by surface weather and equipment problems, the manned submersibles *Alvin, Aluminaut, Cubmarine PC3-B,* and *Deep Jeep* nevertheless scoured the rough

Mud clouds from an undersea avalanche billow around the bathyscaph FNRS-3 *in Toulon Canyon, off southern France. During more than an hour of immobility—before the craft began rising slowly to the surface—the crew feared deep mire had buried them 5,250 feet down. The bathyscaph, carrying Captain Cousteau and Lt. Comdr. Georges S. Houot of the French Navy, touched off the slide after settling on a ledge while exploring the canyon in 1954.*

PAINTING BY ROLF KLEP

147

Lashed by waves that tore loose a towline, the bathyscaph Trieste *undergoes repairs before its 35,800-foot descent in 1960 into the Mariana Trench—earth's deepest known chasm. Minutes later,*

THOMAS J. ABERCROMBIE, NATIONAL GEOGRAPHIC STAFF

Jacques Piccard (right) and U.S. Navy Lt. Don Walsh climbed inside the craft and ventured down nearly seven miles. They touched bottom near a slow-moving fish, proving life exists in the abyss.

ocean bottom until, within 80 days, *Alvin* found the bomb at 2,550 feet.

As Bill Rainnie, *Alvin*'s debonair chief pilot, remarked to me, "With the 1966 Spanish bomb hunt the manned submersible came of age." In short order the deep-water taxis presented their own case.

The Reynolds *Aluminaut*, made of forged aluminum, retrieved a lost string of current meters off the Virgin Islands while conducting an undersea survey and inspection of a Navy tracking range. Scientists at the Woods Hole Oceanographic Institution and the Navy's Electronic Laboratory, after diving in *Alvin, Deepstar-4000*, and the diving saucer, published scientific papers and discoveries amply supporting the case for man-in-the-sea. Lockheed Corporation's *Deep Quest* between January and March 1969 retrieved flight recorders from downed aircraft. The sleek *Deep Quest*, with ex-*Trieste* pilot Larry Shumaker at the controls, followed up with a 3,400-foot salvage of an entire Navy fighter plane that had crashed 26 years earlier off the California coast.

Around the world the undersea vehicles brought into sharp focus the limitations of

Stripped of its outer shell, the diving saucer gets a cleaning and checkup on Calypso's *deck. Power assemblies lie between inner and outer hulls, so the crew faces minimum danger from fires or noxious gases caused by motors or damaged batteries. The hydraulic claw extended above the viewing ports picks up samples from the continental shelves. At right, the saucer hovers for a portrait in the Mediterranean Sea.*

deep-ocean study and salvage operations from surface ships.

In early 1960 I was involved in an ocean-bottom study of the Tongue of the Ocean, a long, deep channel in the Bahamas. On the basis of thousands of photographs and hundreds of miles of soundings from ships, we concluded that the 6,000-foot-deep channel was essentially free of boulders and cliffs below a thousand feet or so. In June 1966 pilot Val Wilson and I made an excursion in *Alvin* over this same "flat and featureless" area. My faith in surface surveying was severely shaken as we slowly picked our way around *Alvin*-sized boulders and over sheer cliffs 200 feet high.

Alvin herself was the object of a search that ended in the deepest and heaviest

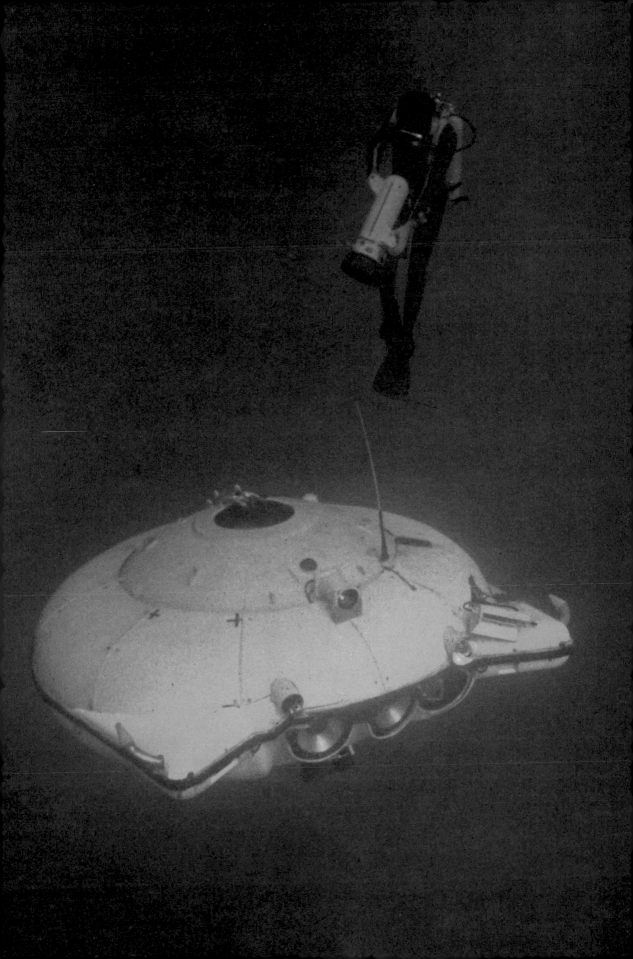

salvage job ever performed. In October 1968 the 15-ton submersible fell from her cradle off Cape Cod and sank—with no men aboard—in 5,000 feet of water. Almost a year later her companion of the Spanish coast search, *Aluminaut*, crawled up her side and attached a heavy line; with a mighty heave the Navy research ship *Mizar* pulled her back to the surface. Less than two years later *Alvin* was diving again.

But *Alvin*'s deep, unscheduled cold storage was not without scientific benefit. The would-be passengers' lunch—six ham sandwiches, two apples, and two bottles of beef broth—were soaked but in almost perfect

condition. This totally unexpected discovery led Dr. Holger Jannasch of Woods Hole to perform further tests on organic materials and confirm the extremely slow rate of decomposition in the deep sea.

Dr. Jannasch explained the implications to me: "A suggestion for disposing of garbage and trash involves baling and sinking it into the deep ocean. Before the *Alvin* findings we had no strong basis for questioning such a solution, but now we have reservations. Slow decomposition means big accumulations of garbage, irretrievable and thus beyond our control. For one thing, gases produced within the bales might

siphonophore as a major component of the DSL's. In the past nets from ships had torn such delicate animals into unrecognizable fragments.

Another biologist, Larry Dunn, a graduate student on leave of absence from the School of Oceanography at the University of Rhode Island, was on active duty with the Navy when he made his first deep dive with me in *Alvin* in 1967.

For me, a geologist, it was an education to learn the names and personalities of the creatures drifting and swimming by our viewports as we descended. We were both amazed to observe the penetration of sunlight as deep as 2,100 feet in the clear Bahamian waters. (Later, in *Ben Franklin,* I was surprised again at being able to read the large print on our charts by sunlight at 600 feet.)

Larry carefully tape-recorded descriptions of the reactions of undersea denizens to our lights and kept track of their numbers and dimensions. Frequently he noted an anomaly in species size or behavior that his textbooks failed to mention. Here were the foundations of ecology, an animal or plant's relationship to its environment. There is an adage of the naturalist: "A fish in an aquarium will do exactly as he pleases." But in the ocean his environment includes those who would eat him: a major influence on his behavior.

At 4,000 feet *Alvin* bottomed gently amidst a tan cloud of sediment. While waiting for the cloud to drift away, we decided to eat. As happens at some point on every dive, I reflected on our science-fiction-like circumstances. Oblivious to the tremendous pressures on our steel bubble, we comfortably ate our sandwiches and chatted about the day's events.

With the sediment cloud gone, we looked

cause them to float to the surface and possibly wash back to shore."

The deep-diving taxis have assembled an impressive record of scientific achievements, particularly for the biologists.

Marine biologist Eric Barham took advantage of the quiet, unobtrusive presence of *Trieste, Deepstar-4000,* and the diving saucer to study the DSL's—the deep scattering layers of itinerant organisms that sometimes play hob with the signals of echo sounders. In addition to contributing a new jellyfish, aptly named *Deepstaria enigmatica,* to our list of undersea inhabitants, Barham identified a fragile, jellyfish-like

out on a soft, light brown sea floor punctuated by small, conical worm mounds. A sea cucumber rose abruptly a foot or so, undulating like an inchworm, then settled down to a sluggish crawl.

"Did you see that?" Larry asked excitedly.

"You mean the cucumber?" I replied, unaware of anything unusual.

"Yes, I can't recall reading anywhere of his moving about like that." A check of his textbooks yielded no mention of this movement. Another bit of information had been added to our paltry knowledge of the deep sea. After the dive Larry admitted that we had probably witnessed other unusual or unknown animal behavior, but that the novelty of a first dive tended to inhibit objective observation. After several dives most people reach a state of relative unconcern, and the mind becomes alert for any new or different sound that might announce trouble or novelty in the normally quiet atmosphere.

Having made a dozen dives or so in other submersibles, I felt reasonably comfortable. When we passed the 3,000-foot level on my first *Aluminaut* dive in October 1967 in the Virgin Islands, I was thoroughly engrossed at the viewport with my nose and forehead pressed hard against the 7.5-inch-thick acrylic plastic. With no advance warning, a sharp crack echoed in my ear. Instinctively, I quickly turned away from the direction of the crack and braced for the rush of cold water. When Bob Canary, *Aluminaut*'s chief pilot, brought his laughter under control, he explained that the water pressure at this level forces the acrylic plastic viewports to seat deeper in their aluminum frames—hence the cracking. Even with this knowledge it was an effort to remain at the viewport, completely unconcerned, and ignore the sound. The reliability and

Silvery bubbles escape from a night diver's breathing gear as he exits from the lock-out chamber of Deep Diver. *Another compartment inside the submersible holds the pilot and an observer. Propellers fore and aft enable the craft to turn 360° within its own length of 22 feet.*
JERRY GREENBERG

Atop Ben Franklin *after a practice dive, the author and Jacques Piccard ready a sack of equipment for transfer to a raft from the support ship. A few days later, on July 14, 1969, the six-man* Franklin *began a successful 30-day test, drifting northward in the Gulf Stream. The Navy's 50-foot DSRV-1, or deep submergence rescue vehicle (opposite, below), seems to stare through a thruster socket at a passing cameraman. Built at a cost of more than 30 million dollars after the loss of the submarine U.S.S.* Thresher *and 129 men, DSRV-1 was launched in 1970; it can lock onto a stricken sub's escape hatch and take aboard 24 passengers. Since 1964* Alvin *(below), workhorse of the submersibles, has taken three-man research teams from Woods Hole to depths as great as 6,000 feet; 1973 modifications rerated it to 12,000.*

FLIP SCHULKE, BLACK STAR

forgiving nature of acrylic plastic has provided major changes in many of the latter-day submersibles.

Perry Oceanographics' *Shelf Diver*, a craft used extensively in pipeline inspection and offshore petroleum activities, added another first in 1968. Transporting personnel to the undersea habitat *Hydro-Lab*, *Shelf Diver* docked onto a tunnel connected to the outside of the habitat and, at atmosphere pressure, provided its passengers "dry transfer." The system, John Perry explained to me, "allows transportation of equipment and personnel to depths beyond that of the ambient-pressure diver, and permits the non-diving specialist economical, physical access to technical or research problems well beyond that of present divers."

Dry transfer is the method used by the Navy's two deep submergence rescue vehicles launched in 1970 and 1971. The 50-foot-long DSRV's can lock onto a stricken submarine as deep as 5,000 feet and, carrying 24 survivors at a time, perform rescues impossible at the time of *Thresher*.

The variety of design and hatch dimensions makes dry rescue all but impossible from the small submersibles. But the designers of each vehicle included a number of options in case of trouble. I was to see some of these options at work in Canada.

I traveled to Vancouver in December 1969 anticipating a dive in *Pisces II*, then under a U. S. Navy contract retrieving torpedoes from the bottom of a test range in Howe Sound. My traveling companion was Mike Costin, an associate from the Naval Oceanographic Office.

Mike and I joined the *Pisces* crew in Nanaimo, and a short boat trip took us to the submersible's bargelike house. Along with Fred Warwick, the young Canadian pilot of *Pisces*, we made ourselves as comfortable as the cold, cramped, six-foot sphere would permit. Towed to the dive site, we checked the essential components and were ready to start our 1,300-foot descent.

Somewhere around 1,000 feet down, while Mike and I were engrossed at the viewports, there was a loud pop, a little

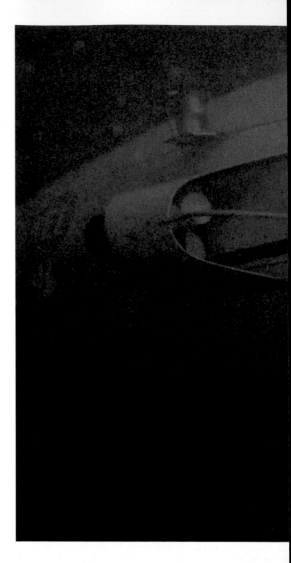

MICK CHURCH (ABOVE) AND NAVAL ELECTRONICS LABORATORY CENTER

Versatile and maneuverable, Deepstar-4000 roams the continental slope near the California coast. The 4000 indicates the working depth of the diving vehicle, patterned after Cousteau's earlier, smaller saucer. The three-man crew — usually a pilot and two scientist-observers — squeeze into a spherical pressure hull 6½ feet in diameter inside the 18-foot-long craft. They communicate by telephone with surface vessels or other submersibles within a range of 7,500 yards. Lead-acid batteries supply propulsion for eight-hour missions, and reversible propellers provide thrust and mobility. A cluster of lights at the nose, for 70mm still and 16mm movie photography, illuminates a world dark since time began. The craft's hydraulically operated claw (opposite) gathers floor samples and deposits them in a collecting basket attached to the hull. An eel-like hagfish slithers out of the way. Brittle stars pattern the bottom beside a deep-water sculpin (lower left) and a rarely seen deep-sea skate (right). Burrowing worms texture the mud.

WESTINGHOUSE UNDERSEAS DIVISION; 1/4 LIFE-SIZE

1/10 LIFE-SIZE

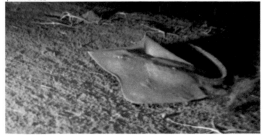

puff of smoke, and an ominous hissing. The electrical wires penetrating the hull from an outside instrument had short-circuited and burned away; the hissing noise was seawater rushing through the holes once occupied by the burnt-out wires.

With no particular tone of urgency, Fred reported to the support ship. "Surface, this is *Pisces*. We just had a minor casualty and are preparing to surface." Back came a laconic "Roger, standing by."

Fred began to blow out water ballast while Mike and I intently watched the depth indicator. Several minutes went by, but the gauge indicated that we were still going down. "Looks like we aren't having much luck," Fred remarked.

The depth was now about 1,200 feet, and I briefly shifted my attention to a coffee mug in the bottom of our sphere; it was covered by an inch or more of water.

With deliberate calm, Fred went through a list of measures he could take.

"Let's see now, we can drop our mechanical arm, or the batteries, or the emergency lead weight," he said, thinking aloud. "I guess we'll try the lead weight first."

At 1,250 feet the 400-pound weight dropped, and *Pisces* floated to the surface.

An incident was to follow ours, however, that precluded the use of such options.

On July 17, 1973, the Smithsonian Institution's *Johnson-Sea-Link* submerged off Key West, Florida, close by a scuttled destroyer at 360 feet. Forward in the acrylic plastic sphere were pilot Jock Menzies and ichthyologist Robert Meek. Aft in the aluminum cylinder were submarine rescue experts Al Stover and Clayton Link, the son of *Sea-Link*'s designer, Edwin A. Link.

Edging close to the listing hulk, an artificial reef where fish gathered, *Sea-Link*

ANDRES PRUNA

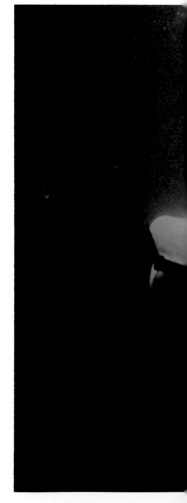

Elevator to the deep, a launch and recovery transport (opposite) carries the two-man submersible Star II. *Towed by ship to a diving site, the transport and its strapped-on sub sink as water fills side tanks, then surface as compressed air blows them clear.* "Forerunner of the future," *the author calls this elevator, developed to avoid the hazards of winching submersibles off and on ship in rough seas. Here, during tests in Hawaiian waters, a diver prepares to strap* Star II *to the transport. A sleek taxi, the* Shark Hunter *(above) serves to whisk divers from undersea work areas to their submersibles. Opposite, above, divers inspect a four-man submersible destined to begin work in offshore oil operations in the North Sea.*

used its forklift to hoist a fish trap left the day before. It slipped off as the sub backed, and on the third try, *Sea-Link* was snared by a steel cable attached to the wreck. The support ship *Sea Diver* called for Coast Guard and Navy help, but hard-hat divers could not descend in the strong surface currents. John Perry's PC-8 tried to find the craft, but its sonar failed and it too was in danger of becoming ensnared.

Thirty hours after entanglement, the situation was desperate. Nothing had been heard from Clay Link or Al Stover for several hours, and it was suspected that the chemicals in their carbon dioxide removal system had expired. Jock Menzies and Robert Meek were weakening and their air supply was almost exhausted.

Then the research ship *A. B. Wood* arrived, carrying an underwater television system. The crew of the *Wood* attached a

grappling hook to the TV apparatus and lowered it to *Sea-Link*. Menzies helped direct the operation by telephone, and finally the hook grabbed *Sea-Link* and pulled it free. Menzies and Meek were spared; but it was too late for Clay Link and Al Stover. They had perished of carbon dioxide poisoning several hours before.

The acceptance of such risks is what each man must face who challenges the sea.

EARLY proponents of the submersible spoke of replacing surface techniques; present-day advocates seek to augment such techniques by providing environmental details and engineering capabilities only man-in-the-sea can offer.

This relationship of mutual support is illustrated by submersible work carried out during the 1970's. Vickers Oceanics' *Pisces II* buried, at 3,000 feet, submarine cable repeaters that were too large for the surface-towed plow to accommodate. Perry's PC-8 was used to provide prospective salvors of a sunken vessel a view of their problem prior to bidding.

Submersibles of the future will apply the lessons learned by *Ben Franklin* and her sister ships. Routine transfer and support of divers and equipment in the offshore oil and construction industries will undoubtedly evolve; periodic inspection of pipelines and hardware will become commonplace, as will pre-salvage inspection of vessels and equipment and even on-site repair.

"But," one might wonder, "you can't work all the time undersea; and with no TV or radio, and when all the books are read, what do you do?"

The answer, if I may return to that successful 1969 Gulf Stream test cruise in *Ben Franklin*, was just outside the viewport.

Hours and days pass quickly when, with only the soft cricket-like chirp of an electronic pinger for company, you lie watching the creatures of the deep ocean passing through the submersible's lights on their never-ending nighttime migration to the surface and their return to the depths at dawn. Long, jeweled chains of barrel-shaped salps gasp in perfect unison as they pump their colony in, seemingly for a close look at the intruder. Tiny, dark copepods, hardly larger than tomato seeds, whirl around and around the bright lights like a swarm of underwater insects. Brilliant red euphausiid shrimp, as if pursued by demons, beseechingly glance at the viewport as they furiously paddle by, trailing antennae almost twice their length. Lightning-fast squid zoom in so quickly and so close that you forget for a moment and jerk back to avoid the collision; but their presence is only temporary, for with nothing but a little dark cloud of ink left as a reminder, they zip off into the dark. Now a pale red, needle-thin worm comes close to the viewport, makes certain he is being watched, and then goes through an incredible series of gymnastics, holding each pose just long enough for the right effect — forming a perfect O or U or S.

Coming into view, the tiny amphipod *Phronima* struggles to drag along a salp, several times larger than himself, that serves as both his food and his home. The shrimp-like little animal makes it to the viewport and, sticking half out of his transparent dwelling, begins to somersault over and over. And every so often, generally just before dawn, the little performers seem to hold their breath and hang motionless as an ominous presence pervades the stage: Without warning, a long, dark, beautifully efficient open-ocean shark glides effortlessly to the viewport lights and casually inspects the fat, yellow and white thing so at home in his domain. But in a few minutes the death machine moves on, and as you sigh your relief, the tiny, darting copepods cautiously start the show again.

Now it is certain: For as long as the show may continue, there will be Piccards and Cousteaus and *Alvins* and *Ben Franklins* to open the curtain and present the cast.

Coppery Caribbean sunset silhouettes the French bathyscaph Archimède, *under tow near its diving site over the Puerto Rico Trench in 1964. Named for the Greek philosopher who discovered the principle of buoyancy,* Archimède *can explore the sea's greatest depths.*

8

Harvests
of the
Future

BY CYNTHIA RIGGS STOERTZ

Resolute ama, *a Japanese diver, readies her mask for a plunge into cold waters for shellfish and edible seaweeds. Her occupation dates back 2,000 years. Today man finds the sea a rich source of fuels and minerals as well as food.*

EIICHI YAMASHITA reached down to an oyster wire dangling beneath the slippery raft and with a delighted grin lifted up a transient sea urchin, a prickly creature about the size of a golf ball. Deftly he cut through the brittle, yellowish-green outer skeleton, exposing the bright orange roe.

"In Japan, sea urchin is a great delicacy," he said, offering me the treat on the tip of his knife. Overcoming a natural aversion to raw seafood, I tasted it. To my surprise, I found it delicious, rather like sweet, fine-grained caviar.

But Mr. Yamashita knows that the American market calls for oysters, not sea urchins. In response he operates the Western Oyster Company, one of the most innovative shellfish farms on Washington's Puget Sound. The oysters that attach themselves to ten-foot-long, shell-strung wires suspended from his rafts total hundreds of times the number that could grow on the bottom in the same area; and they hang well above the reach of starfish, oyster drills, and other predators.

The success of his project depends upon skills Mr. Yamashita has acquired in many fields, including economics, biology, and engineering. His farm typifies a new, scientific approach to aquaculture—the growing of aquatic plants and animals under controlled conditions.

Aquaculture actually goes back 4,000 years or more. A bas-relief from an Egyptian tomb closed before 2,000 B.C. shows a man taking fish from an artificial pond. At about the same time the Chinese learned to grow fish and the Japanese were cultivating oysters. Yet only recently have American aquaculturists and researchers begun to breed marine life scientifically, learn optimum feeding patterns, and study fish diseases.

To see examples of some of the aquaculture that has been undertaken in the United States, I visited Puget Sound, a deep, glaciated inlet of the Pacific that forms a great natural harbor with more than a thousand square miles of water area and 1,400 miles of coastline.

A pioneer of modern aquaculture, Dr.

165

Fantasy of pattern, color, and form, sea shells add a dimension of delicacy and beauty to the marine world. For millenniums, man has valued shells as currency, jewelry, symbols of fertility and good luck, and even as poetic inspiration. Found throughout the world's oceans, sea shells house a diversity of mollusks. Identifications: (1) Venus comb, Murex pecten (4-6 inches); (2) spindle egg shell, Volva volva (3-5 inches); (3) eyed cowrie, Cypraea argus (2-4 inches); (4) tent olive, Oliva porphyria (3-5 inches); (5) perspective sundial, Architectonica perspectiva (1-2¼ inches); (6) pontifical miter, Mitra stictica (1-3 inches); (7) Elliot's volute, Amoria ellioti (2-3 inches); (8) variegated turret, Turritella variegata (2½-4 inches); (9) spindle shell, Fusinus fusus (4-8 inches); (10) banded tulip, Fasciolaria hunteria (2-5 inches); (11) strigate auger, Terebra strigilata (1½-2 inches); (12) textile cone, Conus textile (2-4 inches); (13) striped helmet, Casmaria vibex (1¼-2¾ inches). All come from the collection of nearly 12 million shells at the Smithsonian Institution in Washington, D. C.

NATIONAL GEOGRAPHIC PHOTOGRAPHER VICTOR R. BOSWELL, JR.

Lauren R. Donaldson of the University of Washington, experimented with fish here for more than 30 years before his retirement in 1973. Selectively breeding for large size, rapid growth, and disease resistance, he produced strains of fresh-water salmon and giant trout. His chinook salmon, transplanted to Lake Michigan only a few years ago, spawned a successful new sport fishery in the Great Lakes.

"We'd better get on down to the pond if we want to see the salmon run," Dr. Donaldson said during my visit. Every year students release 200,000 juvenile chinook and silver salmon into the pond, once a water hazard on a golf course. The salmon swim through a drainpipe into Puget Sound and then on to the sea. Two or more years later, in response to a mysterious call only partially understood by scientists, they return home to spawn—fighting their way up the conduit to the artificial pond. The annual arrival of thrashing silvery salmon attracts visitors, schoolchildren, and barking dogs. The students harvest and spawn the fish, saving a portion of the eggs for the university's hatchery and sending the rest to other laboratories.

With Dr. David W. Jamison of the state's Department of Natural Resources, I visited Justin Taylor, whose family has operated oyster farms on Puget Sound for three generations. The still water of the shallow, diked beds reflected the dark evergreens on the surrounding hills and yellow autumn foliage touched by early morning sun. Tiny oysters by the hundred lay scattered on the bottom. Like many oyster growers in Washington, Mr. Taylor supplements his native crop of Olympia oysters by planting seed oysters from Japan.

So far, the rich waters of Puget Sound

THE AUTHOR: *In 1965 Cynthia Riggs Stoertz spent two months aboard the antarctic research ship* Eltanin *collecting specimens for the Smithsonian Institution. Later she returned to Antarctica as a journalist, reaching the South Pole. She worked as oceanography assistant to a congressman, and was managing editor of the* Marine Technology Society Journal. *At present she is editor of* Petroleum Today.

NATIONAL GEOGRAPHIC PHOTOGRAPHER BATES LITTLEHALES
(OPPOSITE AND BOTTOM); CHUCK NICKLIN (BELOW)

In cool waters off Baja California, swaying blades of giant kelp — the largest marine plant — reach past a diver toward the light. Buoyed by spherical gas bladders, kelp fronds may grow two feet a day, and stretch more than 200 feet to form a tangled canopy at the surface. In this dim forest, fishes and mollusks thrive; bottom-dwelling sea urchins graze on the kelp, but sea otters — a species once threatened with extinction by fur hunters — keep their numbers in check in some areas. Above, a diver offers a snack of sea urchin to a frolicking sea otter. In a multi-million-dollar industry commercial harvesters reap kelp for the chemical algin, used in products as diverse as paint and ice cream.

Shallow pools of a pioneer shrimp ranch nurture delicacies by the million beside Japan's Inland Sea. The owner (below) holds a female ready to spawn. Her young will spend their first three weeks in the tile-lined, temperature-controlled hatchery tanks before being moved to the ponds.

have escaped serious pollution. Mr. Taylor and others intend to keep it that way. But as more homes and industrial plants are built nearby, the interests of those concerned with fishing, recreation, and manufacturing or processing begin to conflict. Planning agencies, therefore, are attempting to control the development and use of fragile coastal areas.

The Lummi Indians of northwest Washington encountered this problem of conflicting options for the coastal zone. At a council meeting in 1968, members of the economically depressed tribe considered leasing the reservation's tidelands to heavy industry. At the same meeting, Dr. Wallace Heath of the Oceanic Institute of Hawaii proposed a Lummi aquaculture project.

The tribe hired Dr. Heath as project director and pitched in vigorously, building dikes and ponds, hatcheries and training facilities. The Lummis hope eventually to employ more than 500 Indians full time. Already they raise three million salmon and trout and 20 million oysters a year in their 700-acre sea pond.

Richard Poole, who directs the oyster project, believes Indian-grown seed oysters may in time take over that market. Puget Sound oystermen now buy almost a

billion seed oysters annually from Japan.

The Lummis also experimentally harvested the red seaweed that grows naturally in scattered beds offshore, and the State of Washington now continues the project. Elsewhere, seaweeds are the basis of a multimillion-dollar industry.

Not many Americans realize they eat or use seaweed products daily. Ice cream, cake icing, salad dressing, paint, toothpaste, shaving cream, beer, and hand lotion are among the many diverse products that contain seaweed derivatives such as carrageenin or algin.

Robert D. Wildman of the National Sea Grant Program told me that for centuries the Japanese have eaten seaweed.

"When I visited Japan recently I would see people in train stations buying individual cardboard containers of food with sheets of red seaweed laid neatly on top," he said. "The Japanese eat seaweed with rice and in soups, candies, and other dishes. Today nearly 150,000 acres of Japan's coastal waters are used for growing *Porphyra*, a genus of thin red algae. This is one of the most profitable aquaculture operations in Japan since *Porphyra* sells for nearly $7 a pound, dry weight."

With few exceptions, most aquaculturists still work by trial and error. Much of the information they could use from university research never filters down to them.

"Every benefit we hope to derive from the ocean demands that we put scientific knowledge to practical use," said Dr. Robert B. Abel, director of the Sea Grant Program. "Sea Grant colleges must extend the results of their work into the day-to-day activities of those who live along the coast."

In 1963 Athelstan Spilhaus, then dean of the University of Minnesota's Institute of Technology, originated the Sea Grant concept. "He took his cue from the system of Land Grant colleges established by Congress more than a century ago," Dr. Abel explained. "The Land Grant colleges encourage researchers to tackle practical agricultural problems in conjunction with efforts of others working in the basic sciences. Then, through extension-demonstration agents, research discoveries are quickly put to use by farmers."

Now Sea Grant colleges have begun to carry out such extension work for the benefit of aquaculture—through what Dr. Spilhaus calls "county agents in hip boots."

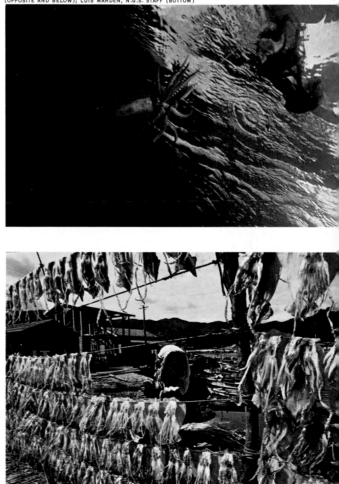

Wielding long-handled dip nets, marine biologists scoop squid—attracted by underwater light—from Africa's Gulf of Guinea. Common life forms of the sea, squid vary in length from 1/2 inch to more than 60 feet. Lured by a bright-colored jigger and then hooked, a short-finned squid (top right) struggles for its life; it squirted the squid-shaped ink blob at right as a decoy. A Japanese worker gathers small sun-dried squid—a staple food for millions. Man also uses squid for fish bait and fertilizer.

Congress passed the Sea Grant College and Program Act in 1966. In its first seven years the Sea Grant Program invested $70.5 million in more than 729 projects in 27 states, the Virgin Islands, Puerto Rico, Samoa, and the District of Columbia.

Increasingly, on all of America's coasts, scientists work with aquaculturists to raise bigger and better crops.

"In Texas the coast does not meet the sea as dramatically as in the Pacific Northwest," said Leatha Miloy of Texas A & M University's Center for Marine Resources. "Beginning about 50 miles inland, the coastal plains slope down gradually, finally merging reluctantly with the warm waters of the Gulf of Mexico. Two hundred thousand acres of lowlands and marshes create a coastal apron that is characteristically both land and sea.

"Because the waters here are warmer, the primary fishery differs from that of northern areas. Texas bays and estuaries are the nursery grounds for a major portion of the nation's shrimp catch."

In the coastal marshlands of southeastern Texas, scientists from the university's Agricultural Extension Service supported by the Sea Grant Program are learning to farm the valued white *(Penaeus setiferus)* and brown shrimp *(P. aztecus)*.

Mrs. Miloy showed me an aerial photograph encompassing the area of the shrimp farms: a patchwork of marshes, small salt water sloughs, and twisting bayous. Bulldozers scooped out five-foot-deep ponds in the clay soil of the marsh. At the end of each pond, gates open at harvest time and a net on the drainage flume traps the shrimp as the water carries them out.

After three years of research, Dr. Jack C. Parker's team believes it has solved major feeding and growth problems. A commercial hatchery operated by Dow Chemical Company provides juvenile shrimp to stock the ponds. Shrimp increase their rate of growth if given additional food, and A & M University researchers working with the Ralston-Purina Company have developed a high-protein feed that remains flaky and appetizing to the finicky crustaceans even after 24 hours in the pond.

"Commercial culture of shrimp is an exciting prospect for increased food production, and it represents a potential use for coastal marshlands—a use that enhances rather than destroys the natural environment," Mrs. Miloy said.

On the island of Martha's Vineyard, the Massachusetts Lobster Hatchery and Research Station conducts studies that may make it possible to farm lobsters. Gulls wheel above the weathered, gray-shingled building overlooking Lagoon Pond and, beyond it, Vineyard Sound. Here John T. Hughes, hatchery director, has spent more than two decades trying to figure out what makes the temperamental crustaceans tick. Several years ago he found that he could get lobsters to breed in captivity by increasing the water depth from 16 to 18 inches in the tanks. Until that discovery the hatchery's chief function was to hold "berried" or egg-bearing females until the eggs hatched, protect the young lobsters from fish, birds, and other lobsters until they could fend for themselves, then release them into the wild.

"We had no way of knowing how many hatchery lobsters survived," he said. "You can't mark them with tags because they shed their shells. Now we can breed distinctively colored ones that lobstermen will recognize as having come from the hatchery."

Among the thousands of lobsters in the hatchery John has a spectrum of colors that includes albino, boiled-lobster red, yellow, green, and bright blue.

"Now that we can raise second and third

Avoiding flailing legs and clutching claws, research biologist Guy C. Powell of the Alaska Department of Fish and Game heads to the surface off Kodiak Island with a batch of five king crabs. Two females dangle from the claws of males, who afford their mates protection during the vulnerable soft-shelled molting and reproductive period. Prized for their succulent meat, mature king crabs may weigh 10 pounds and attain a three-foot leg spread.

Fish ranch of the future: Teams of dolphins, at electronic command from ranchers in submersibles, drive plump fish from a bubble corral to the harvesting area. There suction hoses will draw the fish into ships for cleaning and freezing. Divers control the gate with air valves.

generation lobsters, we can breed them not only for color but also for larger claw size, rapid growth, and other characteristics we want," he explained.

Mr. Hughes held up two lobsters hatched from the same mother. One was four inches long, the other almost a foot.

"The big one grew in water that was consistently about 70° F.," he said. "The small one was raised in water ranging from 35° to 75° F.—a simulation of the lobster's natural habitat. It would take it more than six years to reach one-pound size. This big one took less than three."

He believes lobster farms of the future may use warmed water from power plants to speed growth. "There are a few problems still to work out. Like humans, lobsters enjoy a lobster dinner, so they must be kept apart. We've found that we can cut cannibalism if we keep the lobsters well fed and the water circulating in the rearing tanks. We have tried lobster 'condominiums.' If each has his own apartment, we can keep a large number in one tank. We also may be able to breed lobsters that are not cannibals."

Most of the crops now grown by aquaculturists will end up on gourmet tables; everyday budgets can't afford oysters, shrimp, lobsters, and salmon. But the sea could feed vast numbers of people if the whole range of its food products were put to use. That range includes not only the presently popular seafood but also many kinds of fish now ignored, along with the octopus, squid, eels, sea urchins, and snails for which there is only limited demand, and the seaweed and algae whose potential has hardly been touched.

Chlorella, for example, a genus of protein-rich green algae, could help feed millions. While meat yields 54 pounds of protein per acre, and wheat yields 269 pounds, *Chlorella* yields a fantastic 14,000 pounds of high-quality protein per acre.

Fish protein concentrate (FPC) also could nourish large numbers of people, and someday FPC may be used as a low-cost diet supplement. The National Marine Fisheries Service demonstration plant in Aberdeen, Washington, produced FPC for experimental purposes until its five-year

grant expired in June 1972. Before it closed, John Dyer of the Fisheries Service escorted me around the noisy, immaculate plant which processed hake, anchovies, and menhaden. Inside, I smelled only the pungent odor of alcohol. "The fishy smell is removed during processing," Mr. Dyer said. In an adjoining laboratory he showed me instruments used to check the fluoride and protein content of the FPC. The final product was 80 to 90 percent pure protein.

Food is only part of the sea's value to man. Certain communities in arid seashore areas "desalinate" ocean water for their fresh water supply. Salt has long been obtained from seawater, and now many other chemicals and minerals are extracted, including metallic magnesium.

The President's Commission on Marine Science, Engineering, and Resources has declared, however, that "for the foreseeable future, oil and gas will be the most valuable minerals the nation can obtain from the sea." Not counting production from countries behind the Iron Curtain, offshore or marine wells presently produce about 9.2 million barrels of oil a day—about 22 percent of the world's supply.

But offshore drilling brings new hazards. In 1969 and 1970, three major offshore accidents attracted international attention to the marine oil industry. A well that blew out in the Santa Barbara Channel off California in January 1969 spilled more than 10,000 barrels of crude oil into the Pacific. The oil coated beaches, boats, and waterfowl, and united citizens and local officials in anger and frustration.

A year later, a platform fire in the Gulf of Mexico discharged some 30,000 barrels of oil, and about nine months after that a second fire spilled another 50,000 barrels into Gulf waters.

On the other hand, wells when controlled may actually improve the offshore environment—particularly for sport fishing. The rigs form artificial reefs, and many kinds of fish come to feed on the marine life that grows on them.

THOUGH the oil industry thrives, the future of ocean mining for hard minerals—diamonds, gold, tin—is uncertain.

"There's no doubt about the presence of minerals on the sea floor," Willard Bascom of Ocean Science and Engineering, Inc., told me. "During the mid-1960's when we prospected for diamonds off the coast of Africa, we found deposits of 12 million carats. But it cost more to mine them than they were worth. Storms wrecked or damaged several vessels."

At least one company is more optimistic: Deepsea Ventures, Inc., plans to mine manganese nodules on the ocean floor.

I first saw manganese nodules in the spring of 1965 when I worked with a National Science Foundation research program aboard U.S.N.S. *Eltanin* in the Antarctic Ocean—and realized then the problems miners of the future might face. Our ship rolled in 20-foot seas, and snow lashed across the icy deck. As we winched the huge steel dredge full of dripping nodules out of the sea, the wind and the ship's roll slammed it against the metal hull.

We had lowered the dredge through five miles of black water on steel wire half an inch thick. For our effort, we hauled up only a half ton of the nodules. But we were scientists, and needed only a sample—a commercial firm would need thousands of tons a day. "We plan to dredge a million tons of nodules a year from depths of 16,000 to 20,000 feet, about three to four miles," John Flipse, president of Deepsea Ventures, told me later.

Mr. Flipse believes his company can combine mining with open-sea aquaculture. "Most of the open ocean is barren desert except in areas of upwellings, where nutrient-laden waters circulate from the bottom to the surface. Much of the world's fish catch comes from these areas. As we mine the bottom we produce artificial upwellings that attract schools of fish."

Marine researchers believe tomorrow's harvests will include a wealth of food and minerals; fresh water for large populations piped from undersea springs or distilled from ocean brine; nonpolluting power generated by tides, solar energy, or artificial upwellings; and new wonder drugs extracted from exotic sea creatures—some of them deadly poisonous species.

"The principal thrust of planning for the sea's future harvests will be toward the coastal zone," Dr. Abel told me. "You might almost call it the 'collision zone,' because that's where various interests collide—mining and fishing, recreation and industry, and above all, emotion and practical need. For much of the sea's untapped wealth—harvests of the future—lies there."

Oil production platform stands on stilts in the Gulf of Mexico, linked to the mainland by a radio tower and heliport. Burning waste flares above the installation. Such rigs can draw petroleum from as many as 40 underwater wells.

9

Homes in the Sea

BY CYNTHIA RIGGS STOERTZ

Home for oceanauts, Deep Cabin, a part of Cousteau's Conshelf experiments, rests in the Red Sea 90 feet down. Bottles of breathing gas ring the experimental two-man capsule. Here technicians position floodlights for photographers.

AS THE CLUMSY-LOOKING yellow craft sank slowly toward the sea floor off Hawaii, a diver somersaulted gracefully alongside.

"I felt so free," Danny Kali told me later. "I flew in the water like a bird in air. The umbilical line that carried electrical and telephone cables from the support ship on the surface to our underwater home disappeared into the clear blue infinity above me. Our habitat, Aegir, seemed to glow as it sank into darkness and settled on the ocean floor."

Kali, the skipper of Aegir, and five companions were about to set a world record. For six days in June 1970 the divers worked, slept, and ate 520 feet beneath the ocean surface—almost 100 feet deeper than free divers had dwelt before.

Named for the Norse sea god, Aegir is one of a family of underwater shelters that now enable scientists, explorers, workmen, and rescuers to live and labor in the uncompromising ocean depths.

Aegir's triumph was not easily won. Manned undersea stations have long been envisioned, but building and occupying them had to wait until man devised a way to sustain life—free of a simulated surface atmosphere—far beneath the waves. Ocean engineers in the United States, France, and other nations have experimented for thousands of hours to achieve advances in life-support systems.

But even if undersea dwellings were perfected, could aquanauts safely descend in habitats like Aegir? What about those hazards of deep diving: nitrogen narcosis and decompression sickness (the bends)?

The answers were provided in the late 1950's by Capt. George F. Bond and his U. S. Navy diving pioneers. The Navy team demonstrated that, by using the lighter gas helium to replace nitrogen in the breathing mixture during deep dives, aquanauts can avoid nitrogen narcosis. And once a diver's body becomes "saturated" with the appropriate breathing mixture—compressed air (nitrogen and oxygen) or heliox (helium and oxygen)—he can stay at a given depth as long as he wants without adding to the time he must spend decompressing after

181

his dive. Thus depth alone—not elapsed diving time—determines the length of the decompression period.

"This is the big gain of saturation diving," Bond explained. "It means we can keep a diver down for days, weeks, or even months instead of minutes."

By 1962 two other pioneers, Edwin A. Link and Jacques-Yves Cousteau, each with support from the National Geographic Society, had set up housekeeping at separate locations a hundred underwater miles apart on the continental shelf off southern France. For the first time, man would be more than a casual intruder in the undersea world.

"The ultimate aim," Link told the So-ciety's Committee for Research and Exploration, "is to enable men to live and work on the floor of the ocean at depths of 1,000 feet or possibly more." To many scientists this seemed an unreachable goal.

As the first stage of his long-range experiment, Link launched a submersible decompression chamber in September 1962 from his vessel *Sea Diver* off the French Riviera. Robert Sténuit, a Belgian diver, pulled himself into the cramped, 11-by-3-foot cylinder and descended 200 feet below the surface of the Mediterranean. Then, to leave his pressurized home, he donned his heliox breathing apparatus and slipped through a hatch to swim freely outside—and receive his dinner in a lowered canister. Inside the cylinder that night, he slept fitfully in a sitting position, his head resting on a shelf table.

Just after noon the next day, Lt. Comdr. Robert C. Bornmann, the project physician, ordered Sténuit to winch up to 100 feet. Disappointed because he had hoped for a 48-hour stay, the diver was hauled aboard the mother ship after a little more than 24 hours underwater—the first man to prove in the open sea the feasibility of saturation diving and manned stations.

While decompressing in the cylinder, Sténuit learned the reason for his shortened stay below: A supply boat carrying 15 containers of precious helium he would have needed to continue breathing had sunk in heavy seas on its way to *Sea Diver*.

Only four days after Sténuit emerged from the chamber, Cousteau sent two men, Albert Falco and Claude Wesly, to live 33 feet below the surface off Marseille in Continental Shelf Station (Conshelf) One. Fearing that seasonal storms might buffet

JERRY GREENBERG

Seated in an anchored rubber tube, Floridian Edmond L. Fisher spends a sleepless night during the first endurance test with scuba in July 1954. Every hour for 24 hours, an assistant swam 30 feet down to him with a replacement air tank. At right, Fisher unfolds a kit containing food and fresh water plus a hammer and chisel for collecting coral from Florida's French Reef.

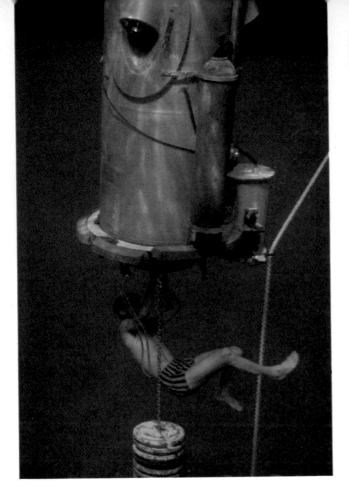

Entering a submersible decompression chamber, Jon Lindbergh, son of the famous flyer, prepares for a descent of 432 feet off Great Stirrup Cay in the Bahamas. The line leads to an inflated rubber dwelling, the SPID, *where underwater pioneers Lindbergh and Robert Sténuit remained for two days beyond the reach of divers from the surface. At right, a hoist lifts the chamber back aboard Edwin A. Link's research vessel* Sea Diver. *Below, Annie Sténuit smiles at her husband through the porthole of a deck decompression chamber.*

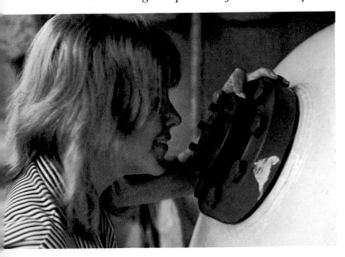

a support ship, Cousteau had set up an elaborate, shore-based operation with a 60-man support team. He strung air, electricity, water, and telephone lines along the sea floor to the habitat, a 17-by-8-foot steel cylinder filled with compressed air, weighted, and anchored to the bottom. Scientists ashore observed Falco and Wesly on television monitors.

The late undersea chronicler James Dugan described the mission: "Each day the oceanauts exited through an open hatch and sortied with Aqua-Lungs for as

long as five hours at a time to a maximum depth of 85 feet. Doctors came down twice a day to examine them, and divers brought mail, newspapers, and hot food sealed in pressure cookers.

"Falco and Wesly had opened the station in high spirits, but on the third day a change came over them. 'I feel small,' Falco wrote in his log. 'I have to go slowly, otherwise I'll never make it. I'm afraid I can't hold out. The work in the water becomes terribly hard.'

"On the fourth day, Cousteau limited visits from the surface and morale improved. 'It's more peaceful now,' Falco wrote. 'I now believe that life under the sea for a long period in greater depths is possible. But what if one should completely forget the earth? When I think about that, I realize I simply don't care what's going on up there. Claude feels the same.'

"Cousteau himself swam down to visit the oceanauts, and observed that concepts of 'inside' and 'outside' were fading away. 'Falco and Wesly passed from air to water and water to air with insouciance, as if the

185

antagonisms of the elements had been abolished,' he wrote later."

When the oceanauts surfaced after 169 hours and 13 minutes, Wesly told Cousteau he was ready to start again. And within a year, in June 1963, he joined six other divers in Conshelf Two.

"This time Cousteau chose a site 36 feet down in the Red Sea, the remote Sha'ab Rumi or Roman Reef, near Port Sudan," Dugan wrote. "Instead of a single ocean dwelling, he designed a small colony—two steel houses, a hangar for the diving saucer, a tool shed, fish pens, and antishark cages. Five men lived for a month in the command center, an assembly of four cylindrical chambers radiating in a design that suggested the structure's name, Starfish House."

For one week of the experiment, two oth-er divers lived 50 feet deeper in the second steel house, Deep Cabin. From its hatch they swam to work daily in 165 feet of water, sometimes diving to 330 feet or more.

Cousteau anticipated that one day trained scientists would be able to study while immersed in the sea, that miners would seek minerals, and that skilled technicians would install or repair underwater oil wellheads.

After Conshelf Two, attention shifted to the Caribbean where Edwin Link and two divers probed the undersea frontier near Great Stirrup Cay in the Bahamas. In June 1964 the veteran Robert Sténuit and marine biologist Jon Lindbergh submerged to 432 feet and remained 48 hours at this lonely depth.

Several innovations marked the accomplishment. The divers were delivered to their deep-lying home by Link's submersi-

Heedless of a swimmer feeding surgeonfish outside, oceanauts play chess in Cousteau's Conshelf Two. The steel Starfish House (below left) sheltered five men for a month on a ledge 36 feet down in the Red Sea. At lower right, divers capture marine life in plastic bags near a toolshed.

ble decompression chamber, used as an elevator. The habitat itself was unique: The sausage-shaped SPID, or submersible portable inflatable dwelling, was constructed of rubber. Flexible walls, Link reasoned, would work as well as rigid ones in a submerged structure where gas pressure inside equals water pressure outside.

Within their heliox-filled home, the aquanauts shivered—despite many layers of wool clothing and an inside temperature of 76° F.—because the helium in their breathing mixture increased the loss of body heat. Through the first night they stood watch and tried to sleep. Next morning they slipped out the open hatch into the sea to observe the teeming marine life, including a curious, scavenging grouper that weighed an estimated 200 pounds. They breathed through hoses 50 feet long

running from the submersible dwelling.

At night sardines flitted beneath the open diving well, chasing plankton in the glare of the light. Without warning, something thudded against SPID. It proved to be the grouper, charging the sardines. The big fish bumped the rubberized station again and again, but caused no evident injury to itself or the tough walls.

After two days and two nights, Link phoned the divers to return to the surface for four long days of decompression, their mission fulfilled.

Two years of exploratory dives by the Link and Cousteau teams served as prelude to the U. S. Navy's first undersea sta-tion. In Sealab I, supervised by Captain Bond, four Navy divers in July 1964 lived and worked for 11 days at a depth of 193 feet in the waters off Bermuda. On sorties from the cylinder that was their temporary home, the men embedded ultrasonic bea-cons in the ocean floor as direction finders; made films, photographs, and recordings; and installed acoustical devices and electric lights to test the effect on sharks. Disap-pointingly, no sharks appeared.

Work was again the password for Sealab II, conducted the following year with Cap-tain Bond as principal investigator. This time the site lay off the coast of California, 205 feet down and a mile from the Scripps

Institution of Oceanography at La Jolla. Twenty-eight divers in three teams occupied Sealab II in 15-day segments; their 45-day total represented the longest manned undersea experiment to that time.

Poisonous scorpionfish, so well camouflaged they blended into the irregular bottom, made sea-floor excursions hazardous. Navy Comdr. M. Scott Carpenter, the world's first astronaut-aquanaut, accidentally touched one of the fish and his hand and arm swelled painfully.

On the first day Carpenter, who headed two of the teams for a total of 30 days, exchanged greetings from Sealab by way of a special radio link with astronaut Gordon

"Papa Topside," U. S. Navy divers affectionately call Capt. George F. Bond, who pioneered the concept that men could live and work undersea.

Hungry almaco jacks gobble bits of sardine fed them from Sealab I, the U. S. Navy's first undersea habitat. In the summer of 1964, four divers lived here almost 11 days, their only connection to the surface a cable for water, power, and communications. At left, the Sealab II habitat—home for aquanauts for 45 days in 1965—bobs off La Jolla, California, before its 205-foot descent.

Checkered sphere of Conshelf Three hangs from cables in Nice Harbor, ready for its first underwater test off nearby Monaco. The steel globe rests on a chassis that holds 77 tons of ballast. Later, it became a home for six oceanauts 328 feet down in the Mediterranean. While living underwater, the men breathed "heliox"—a mixture of about 98 percent helium and 2 percent oxygen. Captain Cousteau (below) tightens a heliox regulator hose in a predive test.

Cooper as his orbiting Gemini 5 spacecraft passed some 150 miles overhead.

Sealab's most popular crew member was Tuffy, a trained dolphin fitted with a harness, who delivered mail and supplies. On one occasion the dolphin dived 200 feet to the house and returned seven times in 20 minutes. Tuffy also demonstrated he could help rescue an aquanaut in distress by carrying safety lines in the turbid water.

In seven years of risky exploration and development, undersea pioneers thus far had sustained no serious mishaps. Encouraged by these successes, the Navy scheduled Sealab III for February 1969 at a depth of 600 feet off the coast of southern California. More than 50 aquanauts trained for this ambitious venture.

Suddenly tragedy struck. In an early test dive, the gear of Berry L. Cannon failed and he died of carbon dioxide asphyxiation. The Navy postponed the Sealab III project indefinitely.

Despite the fate of the deep-water Sealab, a group of scientists determined to push ahead with a thoroughly planned shallow-water study in the Caribbean. On the same day Sealab III would have started, trained

NATIONAL GEOGRAPHIC PHOTOGRAPHER BATES LITTLEHALES

scientists from government and industry—with colleges participating—combined in the joint project Tektite, named for the mysterious meteorite nodules sometimes found on the sea floor as well as on land.

General Electric Company designed and built the underwater house. The U. S. Navy, the National Aeronautics and Space Administration, and the Department of the Interior were sponsors.

"Tektite I allowed scientists to swim out among the undersea creatures to study their living habits firsthand," John Van-Derwalker, a fisheries biologist with the National Marine Fisheries Service, told me.

For two months, VanDerwalker and three other investigators glided through the blue-green depths like big-flippered fish. Since these scientists-in-scuba worked in depths shallower than a hundred feet, they breathed a mixture of nitrogen and oxygen, far less costly than the helium mixture necessary for deeper dives.

The odd-looking habitat—two steel towers atop a rectangular base—sat 50 feet deep on the bottom of Great Lameshur Bay, off St. John in Virgin Islands National Park. In the trees behind the white coral shores a base camp housed a 100-man support crew.

Tektite I proved that the most practical way to study the ocean environment is to become a temporary inhabitant of it. Each scientist, fanning out from the underwater lab, pursued his own specialty. VanDerwalker, observing spiny lobsters, discovered that as he swam close to their dens in the reef they would lash out at him with their antennae. After sunset the lobsters left their dens to feed on sand plains that lay below the reef. VanDerwalker tagged some of the creatures with thumb-size sonic pingers so he could follow them. The plains, he found, are a perilous place for small reef animals: Sharks and other hunters prowl there. He came upon the remains of one of his tagged lobsters; it had been killed and eaten by a predator.

Oceanographer Conrad Mahnken saw brightly colored cleaner shrimp pick and eat parasites from reef fish. "The shrimp seemed fearless, and could easily be enticed to browse along the back of a diver's hand where they assiduously picked at hairs," Mahnken reported.

Dr. H. Edward Clifton, the geologist member of the team, watched sand tilefish build nests near coral heads, first sweeping a depression in the sand with belly and tail and then dropping shells and coral fragments across the top.

Of the four scientists only the team chief, biologist Richard A. Waller, had previous experience in saturation diving. He was a veteran of the Sealab III task force.

A year later, beginning in April 1970, Tektite II evolved into one of the world's most comprehensive underwater research projects. Using the same habitat a second time, a series of five-member teams dived at intervals to occupy the comfortable

Floating on the job, technicians 370 feet down repair a five-ton "Christmas tree," an assembly of tubes and valves that simulates an underwater oil well. Wielding tools made lighter by the water's buoyancy, the divers demonstrated that men can manipulate heavy equipment in the depths.

PHILIPPE COUSTEAU

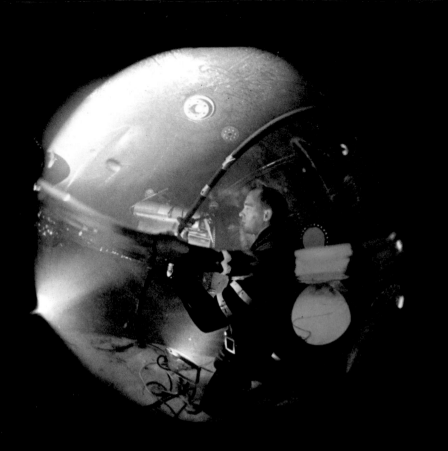

Shoulder-deep in water, a diver kneels on the sea floor inside Link's rubber Igloo, forcing the water out with compressed air from a hose. When inflated, the weighted dome (below) provides a dry work space. The submarino Deep Diver shuttles men 50 feet to the surface near Grand Bahama Island. A submersible portable inflatable dwelling (SPID) at right contains bunks so divers can stay overnight, entering through the bottom hatch.

NATIONAL GEOGRAPHIC PHOTOGRAPHER BATES W. LITTLEHALES (RIGHT) AND JERRY GREENBERG

home on the floor of Great Lameshur Bay. The first crew of women aquanauts participated. Four of the women were scientists; the fifth was an engineer.

As they swam through azure waters, the Tektite II researchers experienced one element of the underwater world for the first time: Using a new breathing system that recirculated the air instead of discharging it in noisy bubbles, swimmers could hear the symphony of sea sounds—the grunts, crackles, and whistles of reef life.

To find out more about Tektite's accommodations, I talked with Mary Kell, a writer and research consultant in the field of commercial deep-sea diving. Mary had swum down to see the habitat between missions. As she described her visit I reflected on the vast difference between this split-level dwelling and the earlier austere habitats.

"The once pure-white Tektite had accumulated a layer of marine growth," she said, "giving it a ghostly greenish cast in the clear water. Long white tracings marked places where parrot fish had nibbled the algae. We swished aside the barracuda lurking in the entrance tunnel, wedged our flippers in among the oxygen bottles, and climbed the ladder into the warm 'wet room.' We toweled off so as not to drip water all over the carpeted floors of the other three chambers. The luxurious habitat seemed much like a mobile home, with all its built-in comforts.

"We were allowed to stay only a little over an hour," Mary continued. "Beyond that time, we would have to undergo decompression. Like the aquanauts, we were observed constantly on TV by the watch director in the command van ashore."

Tektite scientists accomplished more in two weeks of saturation diving than they

Aquanauts relax with music inside Tektite II, undersea headquarters for ten five-man scientific teams in 1970. Fifty feet down in Great Lameshur Bay off St. John in the U. S. Virgin Islands, marine zoologists—members of one Tektite team—test the effects of pollutants on corals; heavy tape holds vials of chemicals.

could have done in several months of surface diving, the participants said.

"Most undersea programs have required
large support teams and therefore have
been relatively expensive," Dr. James Miller told me, "but we were able to hold down
support costs." Then the program manager for Tektite II, he later became deputy
director of the National Oceanic and Atmospheric Administration's Manned Undersea Science and Technology program.

"Of course, a university or small research institute can't afford to put its scientists in the sea on a scale comparable to
Tektite. But look at this." He spread a sheaf
of photographs in front of me. "Here are
some of the low-budget habitats that exist
right now—Hydro-Lab, which Perry
Oceanographics built as an underwater
classroom for Florida Atlantic University;
Edalhab, built by students at the University
of New Hampshire as an engineering project; Aegir, built by a Hawaiian firm; and
Sublimnos, developed by Canadian physician Dr. Joseph MacInnis as a shelter in
cold water for two to four divers."

MacInnis calls Sublimnos the "Volkswa-
gen approach to undersea habitats." Total
cost of this refuge was only about $15,000.

La Chalupa, built by Perry Oceanographics, completed ten manned missions off
Puerto Rico in late 1972 and 1973. This sophisticated and economical undersea laboratory even contains an on-board computer.

The National Oceanic and Atmospheric
Administration has leased some of these
habitats for its programs, Miller told me.
Using Edalhab, NOAA biologists studied fish
behavior off the coast of Florida hoping to
improve commercial catching methods.

*Bits of tinned oysters scattered by engineer Todd
Atkinson attract a grouper and yellow-tailed
snappers to the hatch of Tektite II. Aquanauts
(upper right) spy on fish at long range, using a
camera equipped with a 500-mm telephoto lens.
A modified Ocean-Eye housing, developed by
National Geographic photographer Bates
Littlehales, protects the lens. Marine botanist
Sylvia Earle, head of an all-female Tektite II
team, checks algae growth in a shielded "garden."*

At some locations permanent undersea laboratories may one day enable scientists to carry out long-term research programs. For example, the State of Texas hopes to build an ocean research center 120 miles offshore on Flower Gardens Reef in the Gulf of Mexico.

Other countries are developing sea-bottom laboratories, too. Japan's Seatopia can accommodate four divers up to a month at 300 feet. Helgoland was set up as a permanent underwater station in West Germany. Like Hydro-Lab, its power supply and breathing gas arrive by umbilical line from an unmanned buoy. A Soviet habitat, Chernomor, is used for geological and biological studies.

I TALKED with habitat people in Hawaii, California, Texas, Rhode Island, Puerto Rico—even the Great Lakes. They have plans, some involving international cooperation, for underwater stations that range from simple two-man shelters to large, elaborate structures.

"We hope to have a number of movable, working undersea laboratories," NOAA's Miller said. "We look to the time we can sever all ties with the surface. To do this, we need dependable undersea power sources such as fuel cells, diesel generators, or nuclear power plants. Then man can go to depths limited only by his ability to breathe at increasing pressures."

John VanDerwalker echoed Miller's desire to reduce dependence on the surface. He compared a diver in a habitat to an astronaut far from earth in a spacecraft. "Astronauts take their life-support systems with them," he noted. "So should explorers of inner space. Once divers become self-sufficient, they will be able to work economically at even the deepest reaches of the continental shelf."

Already commercial diving companies operate their own saturation diving chambers in several different parts of the world. Instead of a habitat fixed to the ocean floor, they employ movable undersea stations. Divers live aboard a support ship and take a pressurized bell as an elevator to the bottom. One bell is even equipped with a hot, freshwater shower. When a diver finishes his work, he rides the bell up to a deck decompression chamber. Meals and dry clothing are passed to him through an air lock as his body slowly returns to normal pressure. The decompression chamber is equipped with bunks, toilets, and ceilings high enough to permit a man to walk around.

One "saturation habitat" even features piped-in music, diver-selected radio programs, television entertainment, and wall-to-wall carpeting.

Should a bad storm blow in, the entire vessel—with the divers snug in their chamber—moves to a sheltered harbor. In fact, ship, divers, bell, and chamber can all move whenever the job site changes—which can be frequently.

The versatile Aegir does not require a separate diving bell. It transports divers to the bottom and back, serves as an undersea bunkhouse during an underwater job, returns the divers to the surface, and finally acts as a deck decompression chamber—all for one admission price.

"Entrepreneurs are beginning to think in even more grandiose terms," said Dr. Robert B. Abel, head of NOAA's Sea Grant Program. "Underwater hotels off Santa Catalina, submerged restaurants off Bermuda, a museum in Puget Sound—all are on the drawing boards."

Perhaps on some future day, as armchair aquanauts, we may indeed vacation in an undersea cottage, a tranquil oasis isolated from civilization's pressures. And as we watch the sea's varied inhabitants flicker past our windows, we may recall that only a few years ago brave men first risked these hidden depths to extend the frontier of knowledge, to learn to live in the world beneath the sea.

Twin structures of Tektite II rest on the floor of Great Lameshur Bay. From here in 1970 a series of research teams—including an all-female group—staged an ambitious, months-long assault on the secrets of the sea.

Acknowledgements

The Special Publications Division is grateful to the consultants and authorities named or quoted in the text and to those listed here, for their generous cooperation and assistance during the preparation of this book.

Mrs. James Dugan; Dr. Leonard P. Schultz, Senior Zoologist, Miss Maureen E. Downey, Division of Echinoderms, and Mendel Peterson, Museum of History and Technology, all Smithsonian Institution; Richard Vetter, Executive Secretary, Committee on Oceanography, and Lee M. Hunt, Executive Secretary, Mine Advisory Committee, both National Academy of Sciences.

Dr. Edward Wenk, Jr., Executive Secretary, National Council on Marine Resources and Engineering Development; Haven Emerson, Vice President, Sanford Marine Services; Gordon Kazanjian, Director, Marketing Division, Ocean Systems, Inc.; Lt. Comdr. C. W. Larson II, Public Affairs Officer, and Norman Hanson, Assistant PAO, Deep Submergence Systems Project, U. S. Navy; Russell Greenbaum, Head, Public Affairs, Office of Naval Research; Richard Vahan, Assistant Curator, New England Aquarium.

William S. Glidden, Marine Biologist, and Larry K. Hawkins, Oceanographer, both Naval Oceanographic Office; O. L. Wallis, Aquatic Resources Biologist, and Fred M. Packard, International Specialist, both National Park Service; Comdr. William R. Leibold and Lt. Comdr. John Harter, both Navy Experimental Diving Unit; Richard A. Waller, Biologist, Bureau of Commercial Fisheries; Dr. Russell Keim, Executive Secretary, Committee on Ocean Engineering, National Academy of Engineering; and Ord Alexander, Underwater Engineering Consultant.

Additional References

Oceanography and the sea in general: Robert C. Cowen, *Frontiers of the Sea;* George R. Deacon, ed., *Seas, Maps, and Men;* Rhodes W. Fairbridge, ed., *Encyclopedia of Oceanography;* Sir William Herdman, *Founders of Oceanography and Their Work;* Matthew Fontaine Maury, *The Physical Geography of the Sea and Its Meteorology;* Robert C. Miller, *The Sea;* Claiborne Pell with Harold Goodwin, *Challenge of the Seven Seas;* Sir Charles Thomson, *The Voyage of the Challenger,* Vols. I and II; Hein Wenzel, *The Sea.*

Life in the sea: Rachel Carson, *The Edge of the Sea;* William J. Cromie, *The Living World of the Sea;* Clarence Idyll, *Abyss, The Deep Sea and the Creatures that Live in It;* National Geographic Society, *Wondrous World of Fishes.*

Marine archeology: George F. Bass, *Archaeology Under Water;* Honor Frost, *Under the Mediterranean;* Marion C. Link, *Sea Diver;* Robert Silverberg, *Sunken History;* Joan du Plat Taylor, ed., *Marine Archaeology;* Peter Throckmorton, *The Lost Ships;* Kip Wagner as told to L. B. Taylor, Jr., *Pieces of Eight.*

Diving: Capt. Jacques-Yves Cousteau and James Dugan, *The Living Sea;* Capt. Jacques-Yves Cousteau with Frédéric Dumas, *The Silent World;* Robert H. Davis, *Deep Diving and Submarine Operations;* James Dugan, *Man Under the Sea;* Guy Gilpatric, *The Compleat Goggler;* Pierre de Latil and Jean Rivoire, *Man and the Underwater World;* Owen Lee, *The Complete Illustrated Guide to Snorkel and Deep Diving;* Robert Sténuit, *The Deepest Days.*

Bathyspheres, bathyscaphs, and submarines: William Beebe, *Half Mile Down;* Jacques Piccard and Robert S. Dietz, *Seven Miles Down;* Wilbur Cross, *Challengers of the Deep;* Gardner Soule, *Undersea Frontiers.*

For additional reading on the world beneath the sea, you may wish to refer to the following NATIONAL GEOGRAPHIC articles and to check the Cumulative Index for related material.

George F. Bass: "Underwater Archeology: Key to History's Warehouse," July, 1963. Capt. Jacques-Yves Cousteau: "Working for Weeks on the Sea Floor," Apr., 1966; "At Home in the Sea," Apr., 1964; "Diving Saucer Takes to the Deep," Apr., 1960; "Diving Through an Undersea Avalanche," Apr., 1955; "Fish Men Discover a 2,200-Year Old Greek Ship," Jan., 1954. Dr. Harold E. Edgerton: "Photographing the Sea's Dark Underworld," Apr., 1955. Maurice Ewing: "New Discoveries on the Mid-Atlantic Ridge," Nov., 1949. Anders Franzén: "Ghost From the Depths: the Warship *Vasa*," Jan., 1962. Edwin I. Griffin: "Making Friends with a Killer Whale," Mar., 1966. Lt. Comdr. Georges S. Houot: "Two and a Half Miles Down," July, 1954. Edwin A. Link: "Outpost Under the Ocean," Apr., 1965; "Tomorrow on the Deep Frontier," June, 1964; "Our Man-in-Sea Project," May, 1963. Marion C. Link: "Exploring the Drowned City of Port Royal," Feb., 1960. Luis Marden: "Camera Under the Sea," Feb., 1956. Jacques Piccard: "Man's Deepest Dive," Aug., 1960. Walter A. Starck II: "Marvels of a Coral Realm," Nov., 1966. Robert Sténuit: "The Deepest Days," Apr., 1965. Peter Throckmorton: "Oldest Known Shipwreck Yields Bronze Age Cargo," May, 1962; "Thirty-three Centuries Under the Sea," May, 1960. Kip Wagner: "Drowned Galleons Yield Spanish Gold," Jan., 1965; John G. VanDerwalker: "Tektite II, Science's Window on the Sea," and Sylvia A. Earle: "All-girl Team Tests the Habitat," Aug., 1971.

Index

Illustrations page references appear in *italics.*

Composition for *World Beneath the Sea* by National Geographic's Phototypographic Division, Carl M. Shrader, Chief; Lawrence F. Ludwig, Assistant Chief. Printed and bound by Fawcett Printing Corp., Rockville, Md. Color separations by Beck Engraving Co., Philadelphia, Pa.; Colorgraphics, Inc., Beltsville, Md.; Graphic Color Plate, Inc., Stamford, Conn.; and The J. Wm. Reed Co., Alexandria, Va.